A collaborative project by the staff and participants of Teaching to the Big Ideas

Principal Investigators
Virginia Bastable
Deborah Schifter
Susan Jo Russell

with

Danielle Harrington
Marion Reynolds

Published by
The National Council of Teachers of Mathematics, Inc.
1906 Association Drive, Reston, VA 20191-1502
(703) 620-9840; (800) 235-7566; www.nctm.org
Previously published by Pearson Education, Inc.

Library of Congress Cataloging-in-Publication Data

Names: Schifter, Deborah, author. | Bastable, Virginia, author. |
Russell, Susan Jo, author. | Harrington, Danielle, author. | Reynolds, Marion
(Geometry teacher), author.
Title: Geometry : examining features of shape casebook / principal
investigators, Deborah Schifter, Virginia Bastable, Susan Jo
Russell ; with Danielle Harrington, Marion Reynolds.
Description: Reston, VA : National Council of Teachers of Mathematics, [2017]
| Series: Developing mathematical ideas | "A collaborative project
by the staff and participants of Teaching to the Big Ideas."
Identifiers: LCCN 2017040792 (print) | LCCN 2017050706 (ebook)
| ISBN 9780873539609 | ISBN 9780873539395 (pbk.)
Subjects: LCSH: Geometry--Study and teaching (Elementary)--Case
studies. |
Shapes--Study and teaching (Elementary)--Case studies.
Classification: LCC QA462 (ebook) | LCC QA462 .S3425 2017 (print) |
DDC 372.7/6--dc23
LC record available at https://lccn.loc.gov/2017040792

The National Council of Teachers of Mathematics supports and advocates for the highest-quality mathematics teaching and learning for each and every student.

Printed in the United State of America

Teaching to the Big Ideas

Developing Mathematical Ideas (DMI) was developed as a collaborative project by the staff and participants of Teaching to the Big Ideas, an NSF Teacher Enhancement Project.

PROJECT DIRECTORS Deborah Schifter (EDC), Virginia Bastable (SummerMath for Teachers), and Susan Jo Russell (TERC)

STAFF Jill Bodner Lester (SummerMath for Teachers), Danielle Harrington (Brookline Public Schools), and Marion Reynolds (Tufts University)

PARTICIPANTS Marie Appleby, Allan Arnaboldi, Lisa Bailly, Audrey Barzey, Katie Bloomfield, Nancy Buell, Rose Christiansen, Rebeka Eston, Kimberly Formisano, Connie Henry, Nancy Horowitz, Debbie Jacque, Liliana Klass, Beth Monopoli, Deborah Morrissey, Amy Morse, Deborah Carey O'Brien, Karin Olson, Anne Marie O'Reilly, Janet Pananos, Margie Riddle, Bette Ann Rodzwell, Jan Rook, Karen Schweitzer, Malia Scott, Lisa Seyferth, Margie Singer, Susan Bush Smith, Diane Stafford, Michele Subocz, Liz Sweeney, Pam Szczesny, Jan Szymaszek, Nora Toney, Polly Wagner, and Carol Walker, representing the public schools of Amherst, Boston, Brookline, Lexington, Lincoln, Newton, Northampton, Pelham, Shutesbury, South Hadley, Southampton, Springfield, and Williamsburg, Massachusetts; the Atrium School in Watertown, Massachusetts; the Park School in Brookline, Massachusetts; and the Smith College Campus School in Northampton, Massachusetts

VIDEO DEVELOPMENT David Smith (TERC)

CONSULTANTS Mike Battista (Kent State University), Herb Clemens (University of Utah), Doug Clements (State University of New York at Buffalo), Cliff Konold (University of Massachusetts at Amherst), Rich Lehrer (University of Wisconsin), Gary Martin (Auburn University), Michael Mitchelmore (Macquarie University, Australia), Steve Monk (University of Washington), and Judy Roitman (University of Kansas)

FIELD TEST SITES Boston Public Schools (Massachusetts), Clark County School District (Nevada), Lake Washington School District (Washington), South Hadley Public Schools (Massachusetts), University of Illinois, Chicago, Ventura County School District (California)

This work was supported by the National Science Foundation under Grant Nos. ESI-9254393 and ESI-9731064. Any opinions, findings, conclusions, or recommendations expressed here are those of the authors and do not necessarily reflect the views of the National Science Foundation.

ExxonMobil Additional support was provided by a grant from the ExxonMobil Foundation.

This page is considered an extension of the copyright page.

The following images are from *Investigations in Number, Data, and Space* (Glenview, Ill.: Scott Foresman, 1998) and reprinted by permission of Pearson Education, Inc.: the lettered shapes (from "Shape Cards") appearing throughout cases 1, 8, 18, 21, and in chapter 8 (p. 145), from J. Akers, M. Battista, A. Goodrow, D. Clements, and J. Sarama, *Shapes, Halves, and Symmetry*, a grade 2 unit; one "Find the Block" pattern in case 29 (p. 118) from K. Economopoulos, J. Akers, D. Clements, A. Goodrow, J. Moffet, and J. Sarama, *Mathematical Thinking at Grade 2*; a set of five box nets in case 30 (p. 128) from M. Battista and D. Clements, *Exploring Solids and Boxes*, a grade 3 unit.

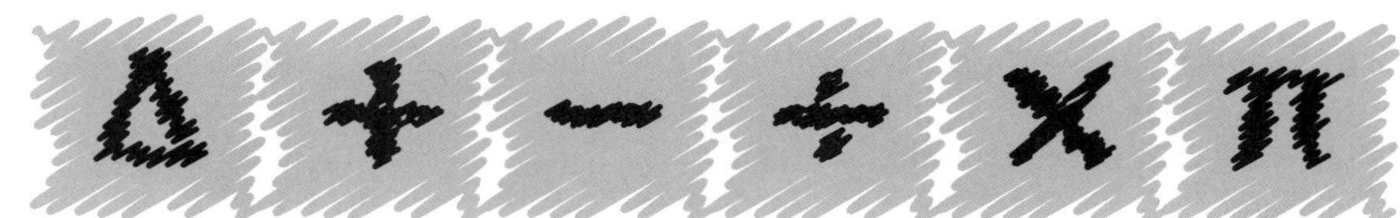

Contents

Chapter 3

Chapter 4

Chapter 5

Introduction

> Through the study of geometry, students will learn about geometric shapes and structures and how to analyze their characteristics and relationships. Spatial visualization—building and manipulating mental representations of two- and three-dimensional objects and perceiving an object from different perspectives—is an important aspect of geometric thinking. . . . Geometry is more than definitions; it is about describing relationships and reasoning.
>
> — from *Principles and Standards for School Mathematics*
> National Council of Teachers of Mathematics, 2000

This statement is as true today as it was when it was written. The study of shape has always been a part of the elementary mathematics curriculum. Depending on grade level, children identify familiar shapes, locate right angles, and examine shapes for symmetry. As the statements from the NCTM document indicate, geometry can also be an opportunity for children to explore the relationships among geometric objects and their component parts.

The cases in the *Examining Features of Shape Casebook* present children's ideas about two-dimensional and three-dimensional objects. They provide a window on the complex interaction between understanding geometric ideas and developing the specialized vocabulary of geometry. Because the cases span kindergarten to grade 7, they illustrate how children's geometric thinking becomes increasingly more complex.

For example, children initially describe shapes by comparing them with objects in their everyday world. Consider the ways these two kindergarten children respond to this shape.

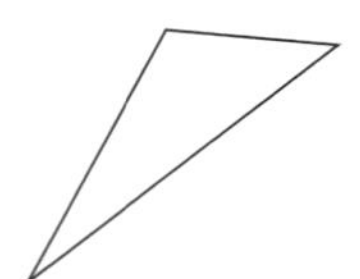

"It looks like a sail of a boat."
"It's like the wing of an airplane."

Both of these kindergartners reacted to the shape as a whole. As children continue to work with shape, they begin to add references to the component parts of the shape to their descriptions. A third-grade student wrote "It has 3 sides, 3 corners, 2 slants, 1 strate [straight] side" to explain how he knows something is a triangle. This child has moved from paying attention to what the shape looks like overall to noticing the components (sides and angles) that make up the figure. The children's thinking in the cases provides insight into the process of moving from general descriptions of objects to formulating and using mathematical definitions.

Children's comments as they work with three-dimensional objects illuminate the mental image they hold for such objects. For example, two third-grade students are talking about a cube. As you read their conversation, work to understand how each child is thinking.

"It takes six squares to make a cube."

"I don't think that is right. It takes much more than that. I mean . . . you would need a whole lot of squares."

Both of these students have correct conceptions for a cube. One is considering the squares that make up the faces of the cube. The other has pictured the cube as a stack of squares, one piled on another. One has a view of the surface of the cube; the other has a view of the cube as a solid object. A full understanding of a three-dimensional object incorporates both of these views. This is an example of one issue children must work through as they begin to deepen their understanding of the relationships between 2-D and 3-D objects.

Through the *Examining Features of Shape* seminar, you will explore key ideas of geometric shape and how children in elementary and middle school come to understand them. The cases were written by elementary and middle school teachers recounting episodes from their own classrooms. All had inclusive classrooms; the range represents schools in urban, suburban, and rural communities. The teacher-authors, who were themselves working to understand the "big ideas" of the elementary- and middle-grade mathematics curriculum, wrote these cases as part of their own process of inquiry. They came together on a regular basis to read and discuss one another's developing work.

The cases are grouped to present children in classrooms that are working on related mathematical issues pertaining to shape. In chapter 1, the cases illustrate children as they describe geometric objects in both two and three dimensions. Chapter 2 provides examples of children developing meaning for geometric terms such as *face, edge,* and *side*. In chapter 3, children encounter and sort out multiple conceptions of angle, and the cases in chapter 4 illustrate children who are developing definitions in the context of three- and four-sided polygons. In chapter 5, children work on the mathematical ideas of similarity and congruence. The relationships between 3-D and 2-D objects are explored through the work of children in chapter 6. Chapter 7 presents instances of children employing geometric reasoning in various contexts: locating the center of a room or of a triangle, determining fractional parts of a square, and finding missing lengths in a diagram.

Chapter 8, the last in this casebook, is the essay "Highlights of Related Research," which summarizes some research findings that touch on issues explored in the cases.

Note: The teachers in this casebook use sets of geometric blocks in their classrooms. One set, pattern blocks, consists of six shapes, each a different color. Although three-dimensional, these blocks are thin so that the emphasis is on the shape of the large two-dimensional faces.

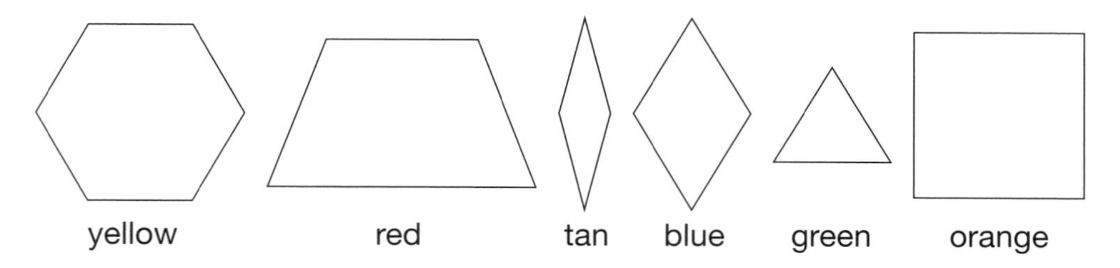

Another set, geoblocks, contains cubes, rectangular prisms, and triangular prisms of varying dimensions, as well as one square pyramid. Examples of geoblocks are pictured throughout the cases.

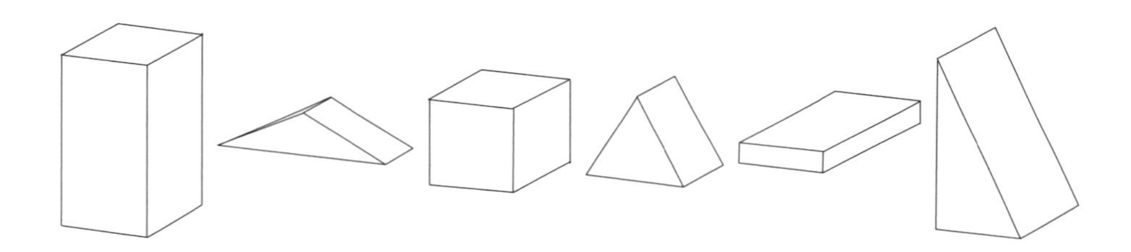

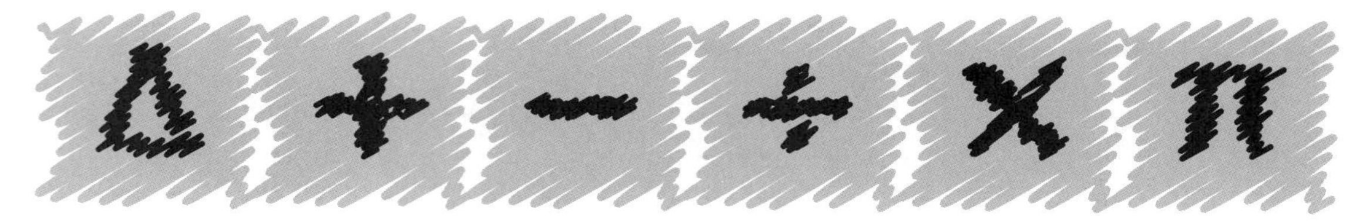

CHAPTER

1

Describing 2-D and 3-D objects

CASE 1
Shapes all around us
Evelyn
Kindergarten, November

CASE 2
Falling triangles
Molly
Grades 1 and 2, September

CASE 3
Describing geometric blocks
Rosemarie,
Grade 1, January

CASE 4
The shape of things
Bella
Grade 1, October

CASE 5
How can it be a triangle and a rectangle?
Alexandra
Kindergarten and grade 1, February

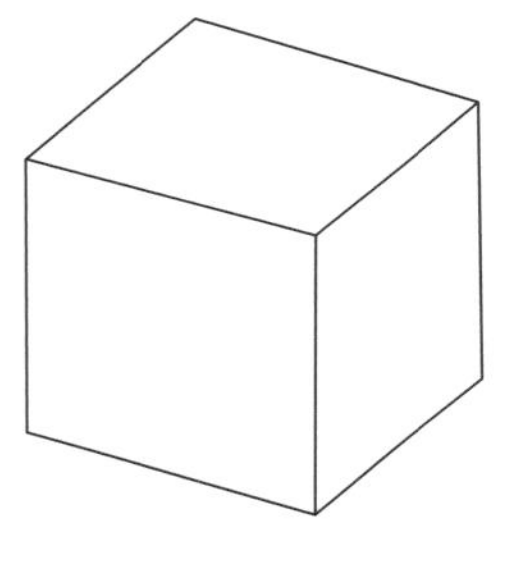

Ask a group of fourth-grade students to describe a geometric shape, such as a cube, and you will hear a variety of ideas. One may tell you, "It looks like a box." Another may notice, "It has six sides and eight points." Yet another may describe it as "a 3-D square." These comments are typical of the ways children describe geometric shapes. They might refer to everyday objects, they might enumerate a list of attributes or components of the shape, or they might comment on the three-dimensional aspect of the object. Sometimes, they demonstrate the way the component parts fit together to make a coherent whole.

Children bring their personal experience with real-world objects to the study of geometric shape. They describe everyday objects by calling upon a variety of attributes, including size, color, shape, orientation, and texture. As children begin to develop a sense of geometric shape, they need to determine which aspects of the object to ignore and which to consider. For instance, a chalkboard eraser and a box of tissues are both rectangular, even though one might be soft and black while the other might be hard and blue.

CASE 6
Observations of geometric solids
Paul
Grade 4, April

While they differ in these and other aspects, geometrically the two shapes can both be seen as examples of rectangular prisms.

In addition to noting which attributes of an object carry geometric meaning, children must also develop the ability to note the attributes of a shape and to coordinate those components into a coherent whole at the same time. This is not an easy task. Consider the reaction of some kindergarten and first-grade children who were asked to describe a triangular prism: "It is a rectangle . . . It is a triangle . . . How can it be a triangle and a rectangle?" They could see both of these shapes, but were not able to reconcile how a single object could contain both. They are just beginning to learn how to coordinate features of shape into a single 3-D object.

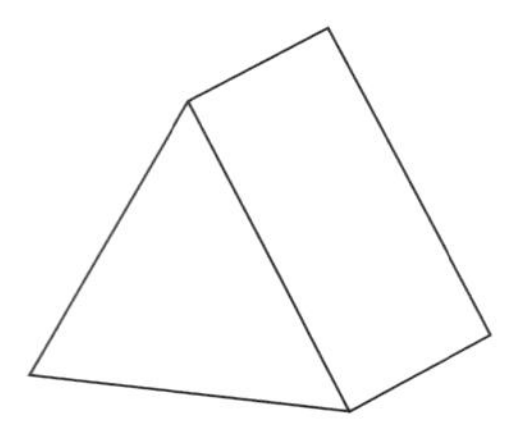

You will encounter these examples of student thinking as you read this chapter. Through the cases, you will visit six classrooms that range from kindergarten to fourth grade, witnessing the ways children describe the shape of geometric objects. The nature of their descriptions provides a window into their thinking about shape. As you read the cases, take note of the following:

- Do the children draw upon their experience with everyday objects or call upon geometric terms?
- Are they thinking of the shape as a whole, or are they focused on a single component of the object?
- Do they pay attention to the orientation of the shape?
- Do they distinguish between objects that are 2-D and those that are 3-D?
- Are they aware that some attributes of the object carry significant mathematical meaning and some do not?

In general, consider what the children's responses might indicate about the way they think about geometric shape.

case 1

Shapes all around us

Evelyn

KINDERGARTEN, NOVEMBER

During a conversation with a third-grade teacher, I became fascinated by her account of some lessons she had done using a set of cards with 2-D drawings. I was wondering how my kindergarten children would react to these shapes. At my request, the third-grade teacher lent me a set of shape cards. Each card in the deck had a different 2-D shape drawn on it. Some of the shapes were quite common: a square, a rectangle, and some familiar-looking triangles.

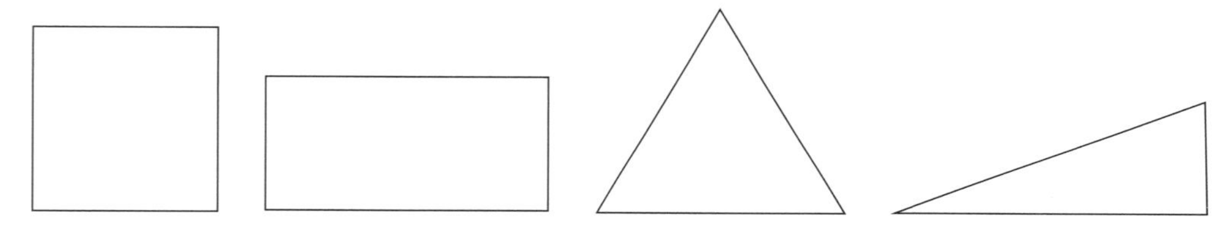

Other cards had shapes that I expected my kindergartners would find unusual. These included a variety of trapezoids and some familiar shapes that I thought would look odd to my children because they were not oriented along the horizontal. I was curious to find out what my students would make of these.

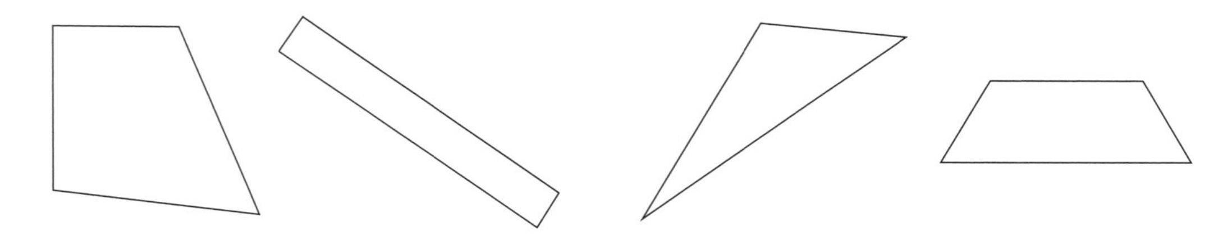

I sat with one group of four students and turned the cards over one at a time. For each card, I asked the children to tell me what the shape was and what they knew about it. The first shape card showed a familiar square.

SANDY: It is a square. It has four sides. They made that language and that's how we speak.

I found it fascinating that she said it was called a "square" because somebody decided to name it that.

JULY: It has four sides. It has four corners.

MITCH: It's sort of, you put two skinny rectangles together; it makes one.

COTY: Square, because if you put two triangles together it makes one.

I was amazed at the variety of their responses. Mitch was describing components of the shape; Coty was visualizing other shapes making this one. We had not done work as a class on these ideas, so it was clear to me that the children were developing these thoughts on their own.

I continued showing the shape cards. The next card had a small square oriented in a diamond position.

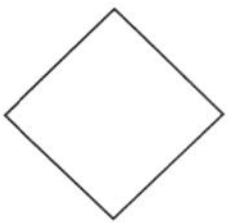

JULY: Diamond. It is like having two triangles and two skinny rectangles.

MITCH: Diamond. If you turn this other one like this [*he takes the square we had just described and rotates it so it is tilted*], it's the same. I mean, the same, but it's a bigger one.

He was comparing the two squares. Even though one was rotated, he noted that both are the same shape. He also commented on their size. The fact that he needed to turn one to make it look like the other didn't make it a different shape for him. He called them both squares.

I decided to show shapes that were likely to be less familiar to them. First, a trapezoid. Their body language changed drastically. Previously, the children were raising their hands and jumping up and down, eager to respond. This time they all sat for a few moments, just staring at the shape.

MITCH: It looks like a crystal. I don't know the name. It looks like a kite.

COTY: Kind of a boot shape. I've seen it on the computer. If you put it like this [*he turns the card a bit*], it looks a little like a diamond.

SANDY: It looks like a kite. If you put a line cross on it, it would be a kite.

Next I showed them a familiar shape, a rectangle, oriented in an unfamiliar position.

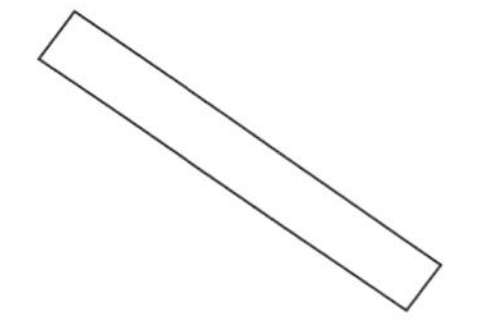

55

JULY: It's a line. If you put another cross on it, It would be an x.

SANDY: It looks like a noodle. You could put it into your mouth and slurp it up!

COTY: It looks like a big piece of Chinese rice. It looks like something you floss your teeth with.

MITCH: It's a skinny rectangle. A really fat noodle. 60

TEACHER: Why is it called that?

JULY: It's a shape of a line.

SANDY: 'Cause it does, just 'cause it does. I ate noodles like that.

COTY: Because there are the points and it's white.

We continued with this triangle. 65

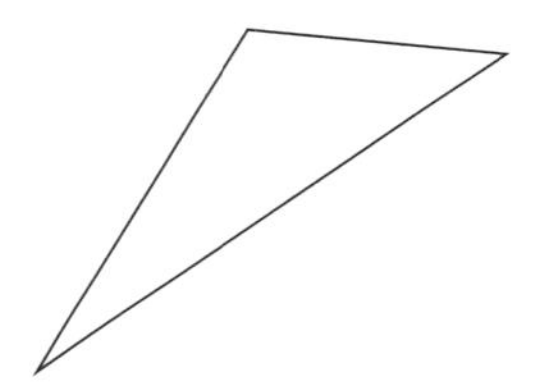

JULY: It looks like a sail of a boat, or a wing of an airplane.

MITCH: It's a crooked wing of an airplane.

COTY: It looks like a hat and sort of a triangle. 70

TEACHER: Why is it called that?

SANDY: It really looks like that.

JULY: I It's a sort of triangle that makes that shape.

COTY: It has a point and three corners.

MITCH: A wing is shaped like that. It's leaning sideways. 75

Finally I showed them this trapezoid.

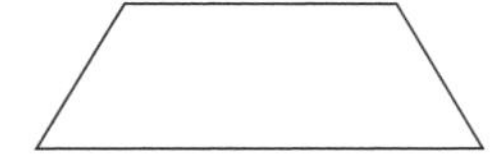

MITCH: It's a dog food bowl.

SANDY: It looks like a rug, a flying rug. It looks like a shape, but I don't know what it would be. 80

As I thought about their responses to these shape cards, I realized that they often made connections between the shapes and objects from their real-world experience. Sometimes they had mathematical language to use. For example, they could identify some shapes as squares or triangles or rectangles. Other times, the children were able to relate to a shape only by connecting it 85 to an object in their life, like a noodle or a piece of dental floss. They also used everyday words, like point or corner, to express their geometric ideas. It will be interesting to see how their ideas and their vocabulary develop as we continue to work with shapes this year.

case 2

Falling triangles

Molly

GRADES 1 AND 2, SEPTEMBER

It's early in the year, and I am doing a lesson in geometry. Today, I asked them to look at an equilateral triangle. "Who can tell me what you know about this shape?" 90

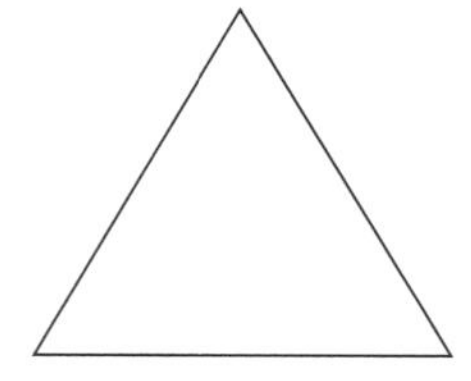

HOLLY:	It's a triangle.	
LIANA:	It has three points.	95
TREVOR:	It's a triangle.	
ALEX:	It has three straight lines.	
NIKKI:	It has like a corner that has a little short top. It has a little corner. [*She comes up to the screen so she can point.*] The corner is small, and it gets bigger and bigger. [*She runs a finger down each side, from the vertex of the top angle, showing us how they get farther apart.*]	100
LIANA:	It has three faces. [*When asked to show us, she points to the three sides.*]	
MELISSA:	It has three flat sides and three corners.	
STEVEN:	It has three points on each side.	

Alex mentioned that if you turn the triangle, the point is crooked on top. As he spoke, he tilted the triangle to the right and left. 105

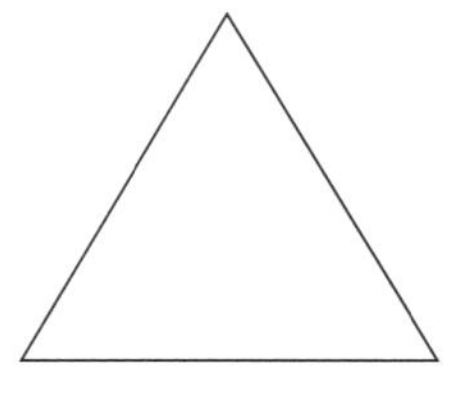

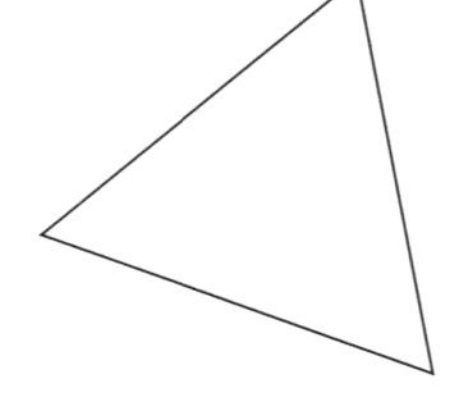

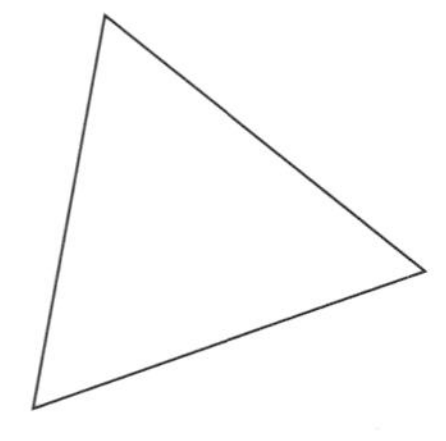

I turned it back so one side was horizontal, and Micah said that, that way it looked just regular. Other children said that it looked perfect the original way, but decided it didn't look right when Alex moved it. 110

I turned the triangle so that it pointed down.

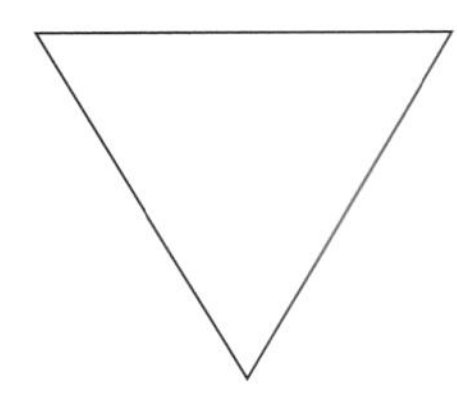

NIKKI:	It looks like it's upside-down because the point is supposed to be up there and it's supposed to come down, and instead the point is down there and it swings around and goes down and it's sort of looking backwards from here.	115

As the children talked, I saw there was general agreement about Nikki's statement. What was it that made this other orientation not correct for them? Was it that they haven't been exposed to other triangle shapes? As I reflect, I think it is more than that. Besides, this wasn't a discussion of what is or isn't a triangle. This was about why a regular triangle shape didn't look right when we changed its orientation. 120

I again asked why it wasn't right. Nikki repeated that the bottom was on the top and the top was on the bottom, but added that if you turn it a little, "it's a little falling to that way." Then Melissa chimed in. 125

MELISSA:	I think that every way it's still right, but the card isn't. It's always the right way. If you're facing this way, it's the right way to you, and [*she turns her body and turns the card*] if you're facing this way, it's the right way to you.	

She was saying that you could always look at it some way and see the bottom flat.

So now I was beginning to wonder, was this the same old discussion about the only "real" triangle being an equilateral triangle sitting on its base? I wanted to find out. I decided to have the class revisit a shape that we had looked at yesterday. It was a right triangle with sides of three different lengths. I asked if there was a right way and a "less right" way for that triangle to sit. There were yeses and nos, and then a lot of spontaneous comments. 130 135

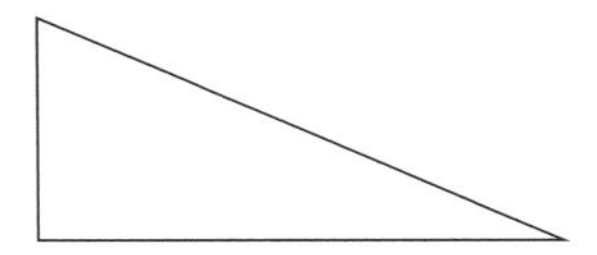

ALEX:	It kind of looks like half a square?	
TEACHER:	It looks like half a square. Is there a right way and a not-right way for it to sit?	
ALEX:	It can sit any way.	
TEACHER:	So how come the first one can only sit this way, but this one can sit any way?	140
ALEX:	No, the other one can sit any way, too.	
TEACHER:	It just doesn't look right? It can sit any way, it just doesn't look right?	

ALEX: Yeah.

TEACHER: Holly?

HOLLY: If it were that way, it couldn't sit. 145

TEACHER: If it were this way, it couldn't sit? What do you mean?

Holly said that it wouldn't feel right, and Alex said that it could sit only three ways, but not on the point. It could only sit on the three flat sides. I rotated the image and showed it "sitting on each flat side" and Alex said that it could sit on those three flat sides and that was what he was trying to tell me. 150

TEACHER: So I have a question about this. [*The image s now sitting on a vertex.*] When you say it can't sit on a point, do you mean that it doesn't feel right, that it looks weird, or do you mean something else by that? What do you mean when you say it can't sit that way? 155

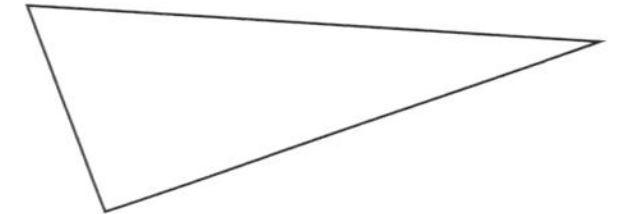

DOMINIC: If you flip it over, it will be the right way. [*He comes up to show me, flipping the triangle so that it sits on a flat side.*]

TEACHER: OK. So Liana, what do you mean when you say it can't sit on a point?

LIANA: It can't sit on a point, because . . . can I go show you something? [*She goes to get a green pattern block.*] If you put this on a flat surface, say like here [*she tries to stand the block on one of the vertices*], it'll tip over. It won't stand up. [*She sets it on one side.*] Right here it will stand up, [*turns it to another side*] right here it will stand up, [*turns it to the third side*] and right here it will stand up. 160 165

TEACHER: So what Liana said is that if we look at this three-dimensional object and try to stand it on a point, it won't stand.

LIANA: It'll just flip over.

TEACHER: So you're looking at this picture and you're thinking about what would happen if it were a thing. 170

Melissa gave us another example by getting a block from the block area—a right isosceles triangular prism—and showing us that this block also wouldn't balance on an edge at the vertex of the face of the triangle.

I then drew an ice-cream cone with a triangle for the cone. I asked if the equilateral triangle could be an ice-cream cone. I heard yeses and a few who said, "No, it's not tall enough." So I asked if our original triangle shape could be the right way now (as an ice-cream cone), and 175 the answer was yes. In fact, Steven added that if I turned it over (put it on a flat side), the ice cream would fall out.

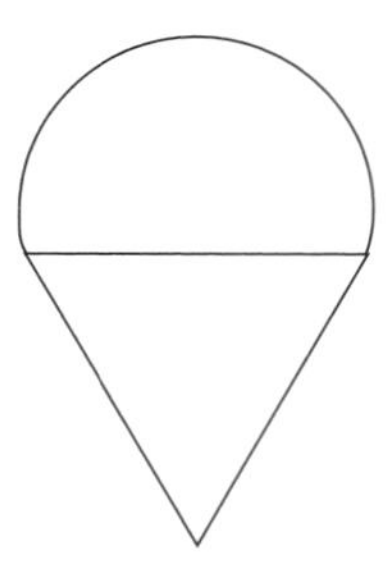

180

Now that this lesson is over, I am reflecting on their comments. Is there something about how children develop these mental pictures that is determined by how three-dimensional objects operate? Maybe a triangle feels "more right" sitting one way because in the physical world it can, in fact, only sit one way. Have my students not yet abstracted these 2-D shapes past their three-dimensional counterparts? Or because the equilateral triangle is the most familiar, does that make the rest not "look right"? I know I'll explore this idea of moving between 2-D and 3-D further this year. 185

case 3

Describing geometric blocks

Rosemarie
GRADE 1, JANUARY

My first graders have been exploring a set of wooden blocks of various shapes. I had noticed during earlier work that they often spoke of the blocks in terms of things they know from their environment; for example, a triangular prism was often described as being "like a ramp." Now I had set up an activity designed to find out explicitly how they would describe the shapes of the blocks.

We started by gathering on the rug. I placed six pairs of blocks in the middle, and asked several children to find matching pairs so that we could make two sets of blocks that were exactly the same. After this was done, I asked a child to put one set in a brown bag. We left the other set on the rug where everyone could see them.

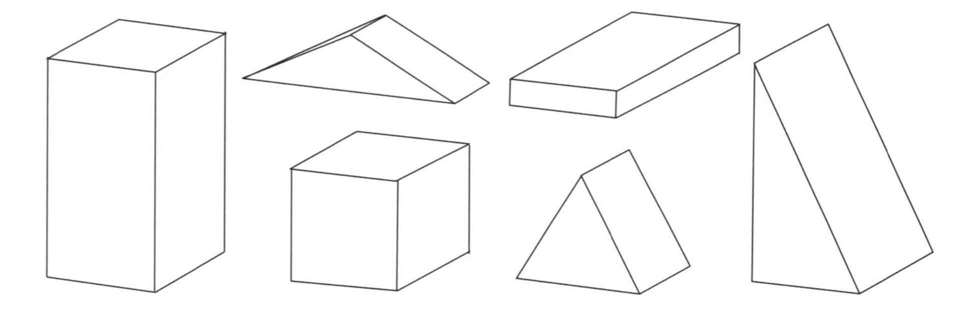

TEACHER: Each block on the floor has a matching block inside the bag. I'm going to ask someone to reach in the bag, pick a block without looking, and describe it—just by feel. Your job is to describe that block really well so that we can find the matching block on the rug, just from your description. For everyone else, your job is to listen carefully and look at the blocks on the rug to see which one fits the description we are hearing.

I then asked Isaac to pick one block inside the bag and describe it without taking it out or looking in the bag.

ISAAC: It has 3 sides. It has 3 points.

EMILY: There is nothing here that has 3 sides or 3 points.

I noticed that both Isaac and Emily were using 2-D terms usually associated with a triangle to describe this 3-D object. Isaac paused, felt the block again in the bag, and changed his description.

ISAAC: It has 5 sides. It has 6 points.

Emily pointed to one of the triangular prisms, and Isaac took the matching block out of the bag. At first, he was describing the block as if it were a triangle, a two-dimensional object. 215
When Emily pointed out that no block matched what he said, Isaac corrected himself. Once Emily chose a block, everyone was satisfied. The other two triangular prisms also fit Isaac's description, but none of the children commented about that. Now it was Gita's turn.

220

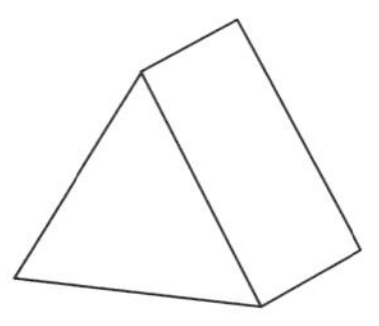

GITA: It has 5 faces, 6 corners. Three of the faces are rectangles.

Gita was calling a "corner" what Isaac had called a "point" and what I know as a "vertex." Just like Isaac, she used a familiar word to express herself. Also just like Isaac, Gita gave a description that wasn't specific enough to pinpoint the particular block. Three blocks in the bag—the three triangular prisms—would all fit the description "5 faces, 6 corners, 3 rectangu- 225
lar faces," but none of the children seemed to notice this. Lakshmi pointed to one of the triangular prisms on the rug, and Gita showed them the block.

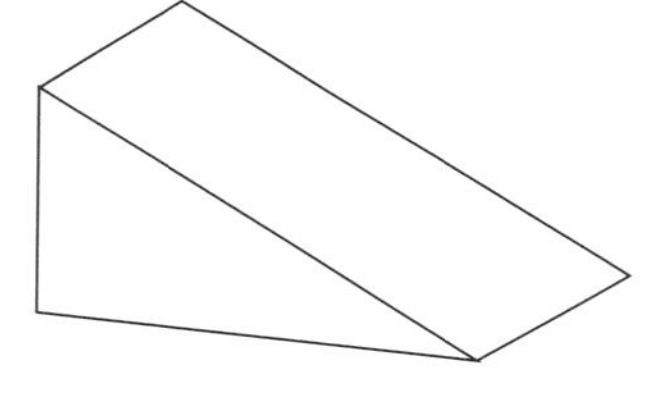

After a few more turns, I asked the class to think about what they had been doing. 230

TEACHER: Tell me what you find easy or hard in this activity.

MELANIE: When Sam said his block had 8 corners and 6 sides, it was confusing. There were two blocks like that. But it was easy when he said it had square sides.

TEACHER: When he said it had 8 corners and 6 sides, it could be any of three blocks: the cube, or the two rectangular prisms. What words did he use that helped you 235
decide which block he was describing?

MELANIE: Sam said that it had squares on opposite sides and it was kind of long.

TEACHER: So you really need to listen to every part of the description before you pick a block.

EMILY: Some kids said that the block has 4 sides, but none of them have 4 sides. 240

Emily's comment reminded me that some children were describing the blocks as if they were two-dimensional, just as Isaac had. Although children are surrounded by 3-D objects in their environment, they often considered only one of the faces for their description. I wonder whether they just talk about the face that appears on the "front" of the block as they look at it, because they can't see what is on the sides or the back of the object all at the same time. Maybe 245 that's why their initial descriptions are 2-D.

After our short discussion, I asked another child to choose a block in the bag by feel and describe it to us.

KJELD: It has 6 sides and 8 corners. It is kind of thick. The sides are kind of long.

TEACHER: [*I see that the terms "sides" and "corners" are becoming commonly used in* 250 *this discussion. I wonder when I should begin to introduce the correct words.*] Is there anything else that you can tell us about it?

KJELD: Two sides are square.

TEACHER: Who can remember Kjeld's descriptions for his block?

SAM: It has 6 sides and 8 corners. It is kind of thick. Two sides are square. 255

KJELD: There is one more thing that I said.

SAM: The sides are kind of long.

I noticed some children were pointing to the thin rectangular prism until Kjeld mentioned the 2 square sides. Chandra pointed out that it has to be the rectangular prism with two square faces. 260

TEACHER: Now I am going to have you do this activity in small groups so that everyone has a chance to describe a block. All the materials you need are in bins on your table. There are six pairs of blocks in a bin. Put one of each pair in the 265 bag.

While the children were doing the activity in small groups, I listened to some of their conversations.

TESSA: OK. It has 5 sides, 6 corners. None of the sides are parallel, and it's like a ramp going down, that's all. 270

Mitchell guessed the block, but Tessa wanted Kjeld to find the block she had described. Kjeld picked the wrong block. She asked Genya, but Genya picked a cube. So then she asked Mitchell, and he picked up this triangular prism:

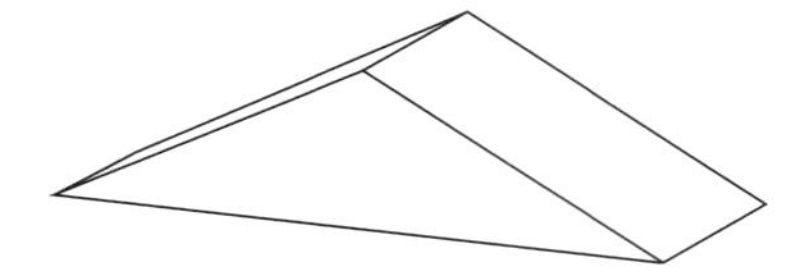

Then Mitchell had a turn to use the bag.

MITCHELL: It has 6 sides and 8 corners. It is kind of long.

TESSA: Is it skinny?

MITCHELL: Yes, it is.

Tessa picked the right block, the thin rectangular prism.

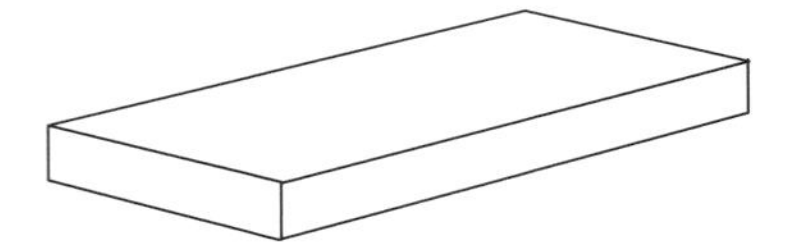

At first, the children tended to make their guesses before they heard all the descriptions. Being able to guess the block quickly seemed to be their focus rather than waiting for more descriptions to be sure they guessed the right block. As the children continued to do this activity, however, they became better at describing the properties of each block. They also began to ask questions if the description was not complete. I also noticed that the children narrowed their choices right away when a block was described as having 5 sides and 6 corners. They seemed to know that it could be any one of the triangular prisms, and they immediately excluded the rectangular prisms.

As I thought about the way the class worked today, I realized that the children were developing vocabulary to describe 3-D shapes. They also were able to identify 3-D objects by matching them to descriptions from their classmates. The ability to give accurate descriptions of the block from feeling it in the bag improved as the other children asked for more information. The children were very much aware of the differences between a triangular prism and a rectangular prism; I saw they noted the number of faces and vertices. By the end of the day, they also knew the difference between the description of a 2-D shape and a 3-D object. I was pleased to see that they were able to choose a 3-D object on the basis of a verbal description. There is more for them to learn—including the meaning of *face* and *vertex*—but this was a good beginning.

case 4

The shape of things

Bella

GRADE 1, OCTOBER

I have a first-grade inclusive class of eleven students. The children were working in groups to explore a set of wooden blocks of various shapes. Mark, Jerome, and Keitha worked together. I watched as Mark picked up a thin triangular prism and examined it very carefully. When I asked what shape he thought it was, Mark said it was a triangle, but it looked like a half of a hill. I asked him why he thought that. Mark placed the triangular solid on the floor to show how the longest side looked like one slope of a hill. He ran his finger up the side as he explained this; then he picked up another block, the same shape and size, and put the short sides together to create a larger triangle, which he identified as a hill. When I asked him what the new shape was, Mark had no trouble telling me that it was a "bigger" triangle. 300 305

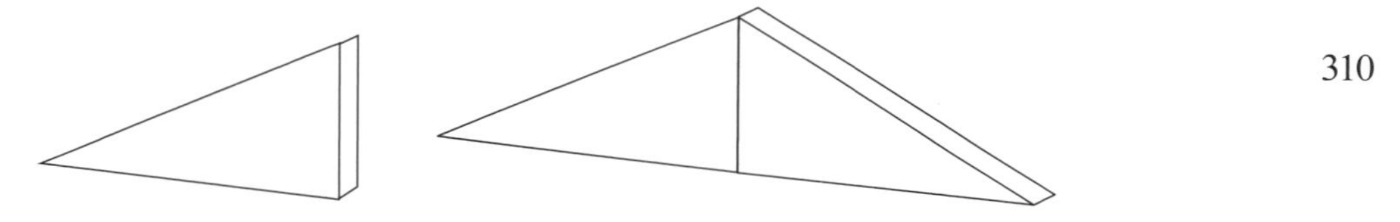

310

Jerome built a structure of four blocks and said it was a gate. He explained that he could drive through the gate and demonstrated this by using another block as a car.

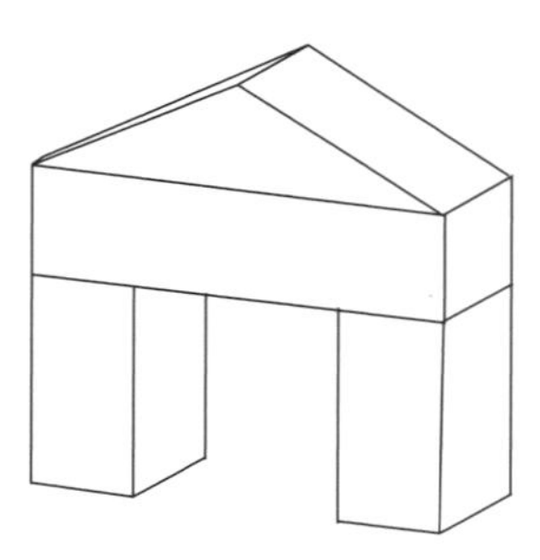

When I asked him what shapes he had used, he called them "rectangles and a triangle," pointing to each block as he named it. I asked him how many of each shape he used, and he said three rectangles and one triangle. 315

Shala caught my attention and I moved over to where her group was working. I asked her to tell me about the large, cumbersome structure that her group had built. She said that it was a house. Nolan chimed in that he thought that it was an apartment building. Shala added a triangular solid to the highest point of the structure, explaining that it was a flag. This surprised me, so I asked her what shape a flag was. She said "a triangle" and pointed out that it looked like our 320 classroom flag. In my mind's eye, I was thinking a flag was a rectangular shape. However, when I looked over to the classroom flag, I saw that it did make a triangular shape as it hung motionless. I listened for a while as Shala described the different block shapes. She had used quite a 325 few cubes in her building, which she called "squares."

As I review what I saw and heard from the children, I am thinking of my next steps. What do I need to do to help them better understand the different shapes? I think about having them look at a specific block, perhaps a triangular block or a cube, to describe what they see.

I also think I want them to compare a cube to a drawing of a square, or a block with trian- 330 gles as faces to a drawing of a triangle. I wonder if this would help them to see any of the other features—like the "3-Dness" of the blocks—which I feel these first graders aren't noticing. They seem to compare the shapes to things that they see every day in their environment. It would be interesting to find out how they would describe a single block without using the environmental descriptors to help them. I'm not sure exactly what they are capable of, but it will be 335 fun trying to find out.

case 5

How can it be a triangle *and* a rectangle?

Alexandra
KINDERGARTEN AND GRADE 1, FEBRUARY

For the past three weeks, we have been using a set of geometric solids for geometry work. Students have had ample opportunity to handle the blocks; many built towers or other structures, and some combined blocks to create other shapes—for instance, joining two triangular prisms to make a rectangular prism. 340

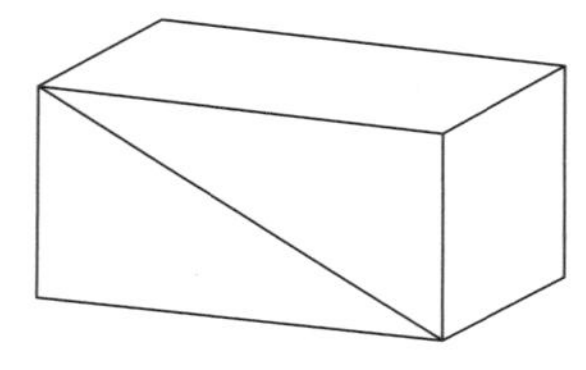

As we began today's lesson, I held up a rectangular prism and kept turning it slowly for all to see. 345

TEACHER: What do you notice about this block?

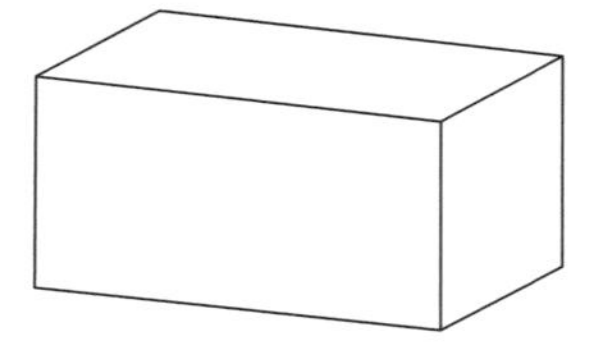

CLAIRE: It is a rectangle. 350

SHANICE: It has a square on the end.

PORTIA: It has six sides.

MEREDITH: It has two long sides and two short sides.

MIKE: It has corners on it, on the end.

JOSH: If you turn it the long way, it looks like my fridge at home. 355

CLAIRE: It looks like a box.

TEACHER:I I would like everyone to come up to the blocks bin and find a block that also looks like a box.

About half the class easily picked out a box-like block; for others it was difficult. All of them looked carefully as they made a choice, offering comments as they picked up and examined each block.

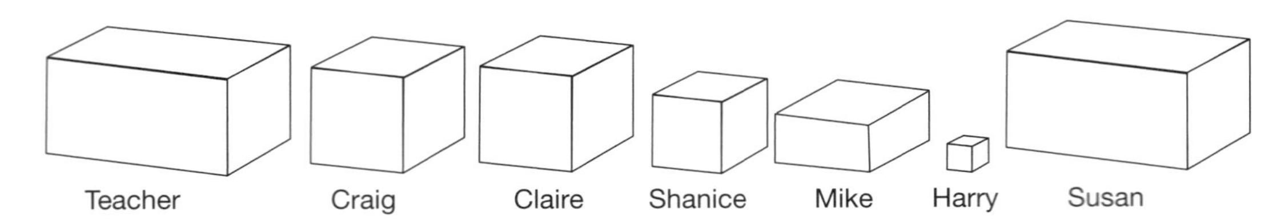

CRAIG: Mine is a square box.

CLAIRE: Mine is too. Two of mine equals yours [*she points to the teacher's block*].

TEACHER: Why do you think that, Claire?

CLAIRE: 'Cause mine is half of yours, so you need two!

SHANICE: Mine is the smaller one.

MIKE: Mine is the flat present.

HARRY: Mine is the smallest one of all.

SUSAN: Mine is the teacher's.

I was amazed to hear these comments. Claire was comparing the blocks, and her second comment about "half" left me speechless. She was seeing her block as part of mine as well as a unit of its own. The ease with which she described a half was impressive. After all agreed that their choices met the one expected criterion ("box-like"), they returned the blocks to the bin.

Then I picked out a triangular prism and asked the class to verbalize what they noticed about this block.

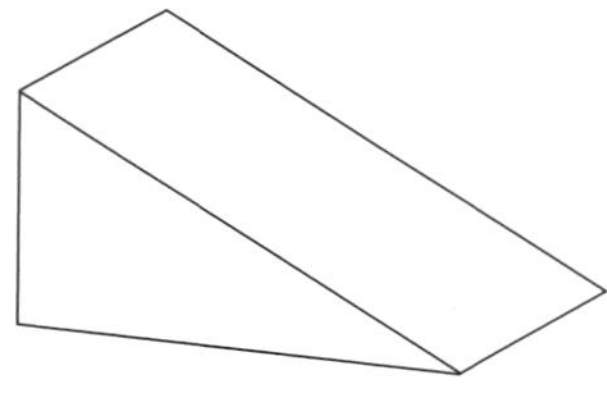

REBECCA: It is a triangle.

HARRY: It has six points.

CHRIS: It looks like a ramp.

PORTIA: This one has five sides.

SHANICE: It looks like a piece of cheese.

MIKE: It kinda looks like a piece of pizza.

MEREDITH: It is a rectangle too, right there on the bottom.

SHANICE: It has a square on it too, the end. It has a triangle, rectangle, and a square.

PORTIA: It has two rectangles and two triangles.

390

For their next choice, the students decided they would look for blocks with both rectangles and triangles. Again, about half the class located a block easily while others needed more time to handle them. These students would select a block from the bin, and then turn it slowly and carefully to examine each face. Several students needed to look at all five faces before making a yes-or-no determination. It was also evident that they viewed each face as "the whole thing" 395 rather than as part of a whole. They could identify a triangular face and later a rectangular face on the same block, but they could not transfer these discoveries to the block as a whole. The idea of a triangular shape and a rectangular shape on one block was confusing. The question, "How can it be a triangle and a rectangle?" was heard more than once.

400

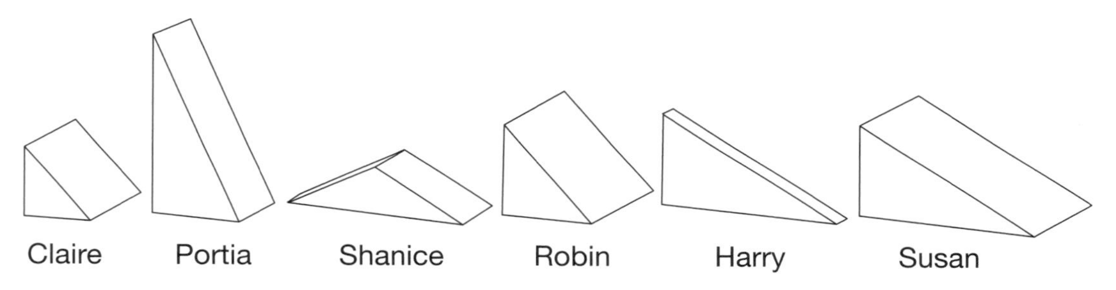

Claire held up her block so we could all see it and traced the different shapes with her finger as she named them. Portia did the same with her block. Shanice held her block in such a way that she showed us only one face at a time. She carefully stated the shape name ("rectangle" or "triangle") as she showed that face to the class. These demonstrations were not sufficient 405 for everyone. It was evident that some students could "see" each face as a triangle or a rectangle, but once the block became a solid, that "seeing" was lost. I asked the children to continue describing their blocks.

SHANICE: Mine has skinny rectangles.

ROBIN: Mine has squares, not rectangles. 410

TEACHER: Why do you say that Robin?

ROBIN: 'Cause the ends are more square than rectangle, so it is square.

HARRY: Mine looks like pizza.

SUSAN: Mine is the same as the teacher's.

I followed up with a related activity that had students focus on the shapes of the faces of the 415 blocks. I gave each student a work mat with six to eight 2-D shapes drawn on it. These served as "footprints" of the faces of the 3-D blocks. To play the footprint game, students worked in groups of six. Each student picked out a block, and then tried to see if it would fill an empty "footprint" on his or her mat. If it wouldn't fill a footprint, the block was offered to others in the group or placed back in the bin. 420

I wondered if the work we did describing and examining the blocks would help the students play this game. For some, the effect was obvious; they were looking at all sides of the blocks to see if any of the five or six faces would fit. Others, however, still seemed to view the blocks as a single flat shape. They never turned the block to view other sides, and only flipped the block over to view the opposite side with my prompting. 425

Some children really made progress as they played. For example, I watched as Claire, Portia, Shanice, Craig, Robin, and Jason got started. They were all excited to play. As each student picked a block, the others scanned their mats to find a fit. They were working as a team; each student was helping to fill all six mats. Both Portia and Claire had a work mat with eight triangle outlines. At first they thought it would be impossible to find eight different triangles, so they 430 were thrilled when they were able to fill their mats. Clearly these six were viewing the geometric solids as just that, 3-D pieces. The ease with which they handled the solids was amazing. All six could "see" all of the faces as possibilities for filling spaces. This was not the case for the students in the class who were focused on only one view. They have yet to learn that the blocks present different 2-D shapes on different faces. 435

Today's work was only the beginning. We will explore together the language that we use to describe the blocks. They already know words like *rectangle* and *square*. From their conversation today, I see we'll have to sort out what they mean by words like *side*, *face*, *corner*, and *edge*. There is a lot of geometry in their future.

case 6

Observations of geometric solids

Paul
GRADE 4, APRIL

440

As an introduction to a geometry unit, I asked the question, "What is geometry?" About half the class conveyed something that related to shapes, but nothing (or very little) beyond that. With this information in mind, I was very curious about their intuitive responses to shapes. What geometric thinking do they bring with them from "everyday" life, I wondered.

I presented my fourth graders with a bag of geometric solids and asked the students to (445) describe the different shapes to me. I wanted to see if they would describe them by using their experience with everyday objects, by using the geometric names such as cone or prism, or by referring to concepts such as faces, edges, and corners. At different points in the conversations, I gave them the geometric name for each solid. Following are some of the descriptions I heard.

450

Cone

"A triangle with a flat bottom."
"A round triangle."
"A large circle with smaller and smaller circles on top until it reaches a point."
"Big to small circles." (455)
"Like a loudspeaker."
"A cylinder, triangle, and circle in one."

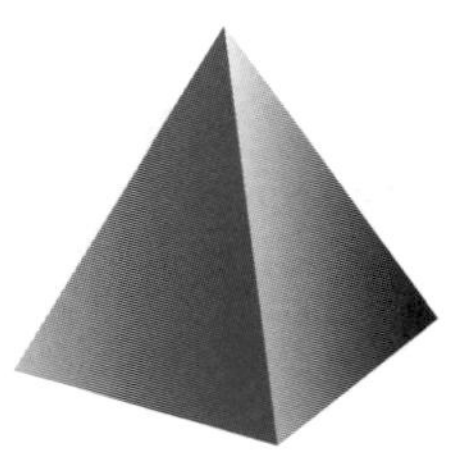

Square pyramid

"A 3-D triangle."
"A cone, but square." (460)

"The top of a house."
"Squares piled on top of each other, getting smaller and smaller."
"A beam of light."
"Flat sides, flat bottom, making a point."
"A cone is a pyramid's cousin." 

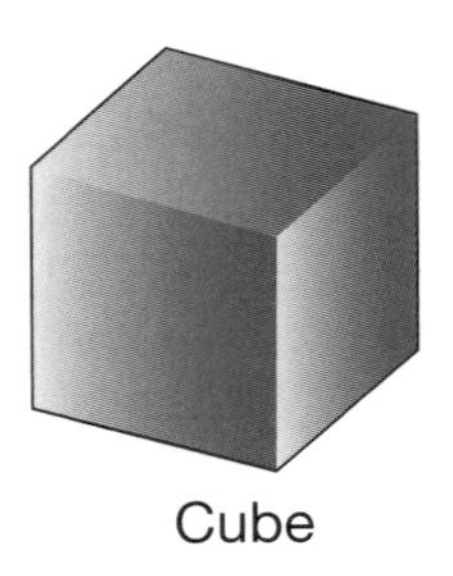

Cube

"Looks like an ice cube."
"It's a box."
"It's a square."
"It has six sides and eight points." 470
"It's a 3-D square, that means that if you stick it flat on a wall, part of it will come off the wall."

Sphere

"A ball."
"A circle."
"Different-size circles piled up." 475

As I considered my students' responses, I was particularly interested in the way they moved back and forth between 2-D and 3-D aspects of the shape. They seemed able to view the 3-D shape as a whole. In particular, I thought that the view of a sphere as "different-size circles piled up" and the comment about the cube as "coming off the wall" were quite wonderful. 480

CHAPTER

2

Developing meaning for geometric terms

CASE 7
Speaking geometrically
Mary
Grades 1 and 2, February

CASE 8
What is a point?
Mary
Grades 1 and 2, January

CASE 9
What is a "corner" really?
Molly
Grades 1 and 2, September

CASE 10
Different ways of
seeing edges
Isabelle
Grade 2, December

CASE 11
Taking a second look
Janine
Grade 4, November

In chapter 1, we examined the ways children described 2-D and 3-D objects by analyzing the language they used to express their ideas. In this chapter, we will focus on the way children develop geometric meaning for specific terms. They begin by borrowing words from everyday life. They call a vertex a *point* or a *corner*, they consider a tilted square to be a *diamond* or a *kite*, and they say *side* or *edge* to mean a face of a 3-D object.

As children discuss their ideas with their classmates, they encounter the need for a shared vocabulary. That is, in order to have a conversation about the number of points or the kinds of faces an object has, the class must first agree on the meaning of *point* (often used by children for *vertex*) or *face*. This is not as simple as it sounds. For example, when the first and second graders in Mary's class consider the question, "Does this shape have three or four points?" a lively discussion of what exactly a point is ensues. Does it matter if it points inside or outside the shape?

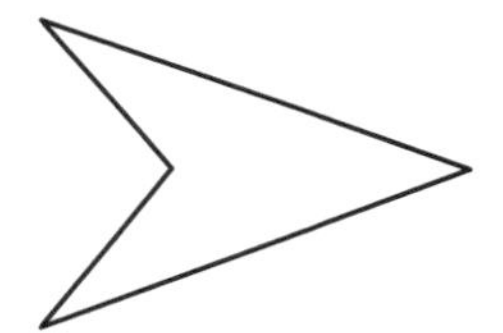

In Molly's class, other first and second graders are working to explain what they mean when they use the word *corner*.

As students debate such issues, they simultaneously build a shared vocabulary and deepen their understanding of the geometric objects being discussed. Through interactions with their teachers and texts, students become aware that the objects they are discussing have been defined by mathematicians. They begin to connect their own language to the more formal geometric terms.

As you read the cases in this chapter, ask yourself these questions:

- What ideas about language do children call upon as they develop a geometrical vocabulary?
- What is the process through which children connect their own meanings for terms with those of their classmates? With those of the established mathematical community?
- What is the interaction between acquiring geometric vocabulary and developing geometric ideas?

case 7

Speaking geometrically

Mary

GRADES 1 AND 2, FEBRUARY

I teach a first- and second-grade combination class. Three days a week, the children receive separate math instruction according to grade level. For the other two days, everyone has math lessons together. The entire class is now in the midst of a geometry unit. They have done activities such as filling in shapes with pattern blocks and making quilt designs with squares and triangles. We have also used 3-D wooden blocks.

As the children work, I assess their understanding and vocabulary. Some describe the shapes in relation to other shapes they know; for instance, they might call a triangle "half of the blue [rhombus-shaped] pattern block." I have found that moving the shapes or having children look at them from different perspectives seems to increase their observations. With the 3-D blocks, some children use language that shows an attempt to describe the elements that make the blocks three-dimensional, while others use only 2-D language to describe their shapes.

One day, I had the whole class build structures with the blocks. They worked in pairs, and drew and wrote about what they had built, from both an ant's-eye view and a bird's-eye view. The buildings, drawings, and teamwork were great, but their language was limited. They simply listed terms like squares, rectangles, triangles, and 5 points rather than write anything longer. This was true for both first and second graders. I wondered what would elicit more spontaneous writing and dialogue. How could I get students to elaborate and explain? I don't usually have such a quiet class! I decided try something else.

I gave the children each a sheet of paper and had them fold it to make two columns. Then I displayed two large blocks, an arch and a rectangular solid. I made a point of turning them different ways so the children would think about them from different perspectives. First we looked at the arch.

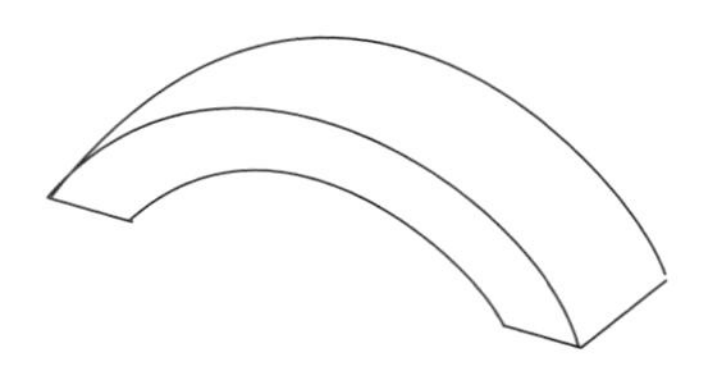

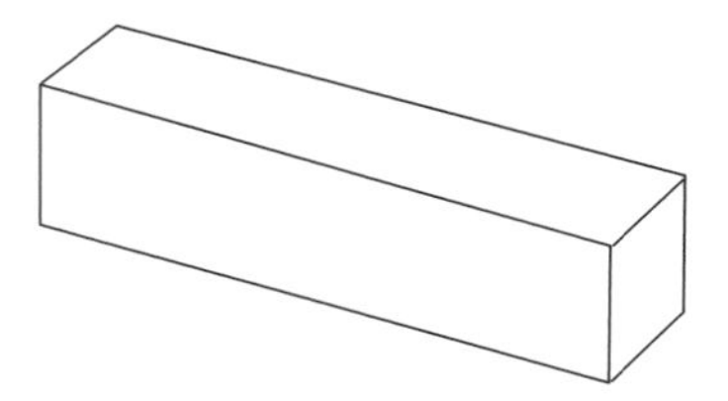

TEACHER: Draw a sketch of this block [the arch] at the top of the first column. Look closely at this block and think about the shape. What is something in our world that reminds you of the shape of this block? Write down everything that comes to mind.

I gave the children plenty of writing time. They liked the activity. As I wandered around the room, I was delighted with the quantity of writing. No one appeared to be stuck. After they had worked by themselves for a while, they shared their lists with the people at their table. If they heard an idea they liked, they knew they could add it to their own list. This is a familiar strategy for my classroom because it creates a situation where everyone has something to say.

They had a lot to say. I recorded their comments on chart paper:

telephone	the letter C	the letter f
roller coaster	smiling mouth	the letter u
rainbow	a road	a slide
sunset	half circle	a house
wedge	arrow	the curves on South America
half of #9	rhino horn	hair
half of zero	half of three	an elephant tusk
part of 5	a beard	a chair

I congratulated them on their work and pointed out that they could always use these sorts of words when they are describing shapes in writing or during discussions.

Next I held up the rectangular block and told them to follow the same format: First draw the block (at the top of the second column), and then record their ideas. After they had shared in small groups, we worked as a whole class and generated another long list:

a tree trunk	a road	a building
a ladder	a stick	the number 1
a stripe	chalkboard	top of a deck
a new pencil	a street line	a log
a flag	lowercase L(l)	the Empire State Building
a box	a bed	an eraser
wood	lowercase I(i)	a marker
a tissue box		

Calling everyone to the carpeted area of the room, I decided to end the lesson by focusing on the geometric attributes that define a shape. When the group was settled, I held up the rectangular block and a tissue box.

TEACHER: When we were writing our lists, some of you told me that this block is like a tissue box. What attributes do they have in common? How are they the same?

The children had a lot to say. My goal was to help them see that certain features are more important in mathematics than others. When they mentioned an important attribute, I wrote it on chart paper. When students started talking about edges, faces, and corners, we counted them. I condensed their ideas to the following list:

similar size

3-D

12 edges

8 corners

6 faces

rectangle faces

In conclusion, I realize that I have two instructional goals for mathematical language. I want my students to be spontaneous and descriptive, but I also want them to use correct and precise language. Today's lesson was to help them see that a real-world object can be considered an example of a rectangular solid. The lesson moved from describing a shape in common terms to describing that same shape with geometric language and to noting that two objects can be seen as having the same shape when they share the same number of faces, edges, and corners. I like the way we are progressing toward a more technical vocabulary. Some children have begun to use *face* and *edge* in a meaningful way; in the future we will talk about what they mean by *corner*. Maybe I'll introduce the word *vertex* then. I'll be watching the students (especially the second graders) to see how these two skills, describing and defining, support each other as we build mathematical definitions.

case 8

What is a point?

Mary
GRADES 1 AND 2, JANUARY

In my first- and second-grade combination class, the second graders were just beginning a geometry unit. For one activity, students worked in groups with a set of shape cards. (Each card shows a different two-dimensional geometric shape.)

I explained the activity: Each group was to sort the cards according to some rule they chose, and then display their work on a poster. As they got started, I went around the class and asked groups about their favorite methods for sorting. Some students sorted by size, but I also noticed that sorting by the number of vertices was very common. (Students were using the word *point* for *vertex*.) I was, however, surprised by some of their decisions. For instance, one team had sorted the following shape (labeled on the cards with the letter L) into a group they called "shapes that have five points." They counted the points this way:

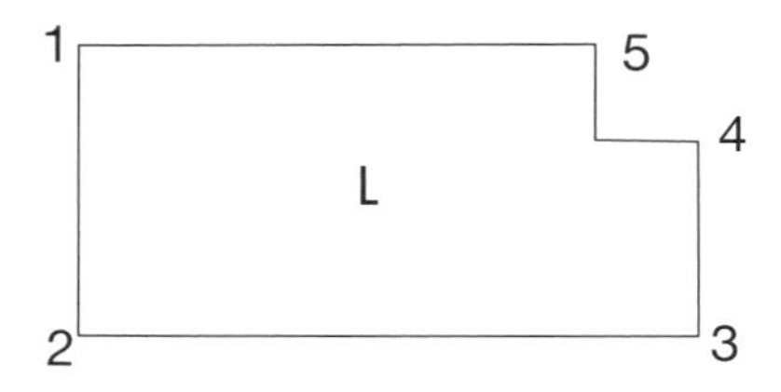

I asked them about the point they didn't count. One told me, "That's not a point because it's on the inside." Another member of the group elaborated, "It's not a point because it goes in. It isn't sharp like the others."

As I spoke with more groups, I quickly found that there was disagreement about two shapes: the shape pictured above (L), and a chevron shape, labeled K. I decided to focus on this issue for our whole class discussion. When the children gathered on the carpet area, I held up the cards for these two figures.

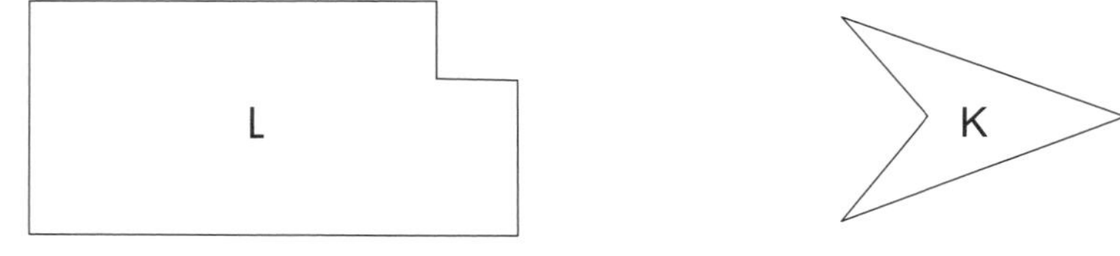

TEACHER: These two shapes are raising some questions in my mind. When I was watching you sort the shapes using points, I noticed that you sorted them in different ways. I'm interested in how you are thinking about the word *point*. First, look at shape K. How many points would you say it has?

JULIAN: Three. [*He counts the obvious points.*]

TEACHER: Who agrees with Julian? [*Over half of the group agrees.*] Why are you not counting this? [*I point to the fourth vertex of K.*] 95

MIGUEL: It's not sticking up.

LENNY: It's not on the outside.

I decided to try something. I took a blank sheet of paper and traced part of shape K as the children watched. I was thinking this mysterious point might seem less mysterious if it was neither inside nor outside a shape. They looked at what I had drawn. 100

TEACHER: Does that drawing change your minds?

LENNY: No.

KYLE: It's a point. It's sharp.

JULIAN: When it's pointing inside instead of outside, it doesn't matter. 105

Chloe's eyes grew big when I drew that part of shape K, so I asked her what she was thinking.

CHLOE: Now I think the shape has four points.

I suggested we look again at the other shape now, shape L. Jenny volunteered that this shape has six points, and she counted them, including the "inside" corner. She had actually 110 counted all the points before the discussion, but she wasn't able to explain why to anyone else. It just seemed to be right to her. And now some of the other students agreed.

DAVID: Six points. This one digs in a bit, but it is still a corner.

However, Lenny and Joel were still quite adamant that "inside points" did not count, and said that shape I had only five points. 115

Thinking back, it occurs to me that if all of their experience has been with convex closed figures, it isn't surprising that the non-convex shapes raise new issues. At first, only two students saw figure K as having four points. My demonstration (drawing just a part of that figure) gave the class something to think about. We will look at more figures with this issue in mind.

I am also wondering when it will be useful for them to consider that a line segment is 120 made up of many points and that the special points they have been counting are actually called vertices. I look forward to watching the growth in their vocabulary and understanding as this unit proceeds.

case 9

What is a "corner" really?

Molly
GRADES 1 AND 2, SEPTEMBER

While planning a geometry investigation for my first and second graders, I noted that the teacher's guide recommended first giving students the definition of a *face* (the flat side of a block). I was reminded of past discussions with other teachers, trying to figure out when it was helpful to provide a definition and when it was better to let children muck about and work out their own definitions. Now I was wondering about the effects of giving or not giving this definition. If I didn't define *face* for my students, would I be withholding information, or would I be pushing them to build their understanding? I wasn't sure, but I did decide to follow my instincts instead of the guide. I decided we would look at a variety of blocks, some from a set of solid wood blocks called *Geoblocks* and some from the set of pattern blocks (which are flatter and thus seem more two-dimensional), to talk about their similarities and differences.

I put some of each kind of block into the center of our circle and asked the children to tell me what was the same about the blocks. They came up with a list that I thought was quite predictable: they're all blocks; they're all made out of wood; they have shapes; you can build with them. Then Nikki said, "The pattern blocks are sort of shaped in different shapes than the Geoblocks." She proceeded to point to the blocks that she saw as different.

Trevor interrupted when Nikki said there was no Geoblock that was the same as the square orange pattern block. He pointed to a rectangular prism that looked somewhat like the orange pattern block but was larger. As they discussed the differences, they pointed out that the pattern block looked "flatter" and the other was "sort of bigger, and stuff."

I then set out the two blocks that Nikki and Trevor had been discussing and asked the children to describe them. Both were rectangular prisms; one was thinner than the other. I asked the students to tell me how they were the same or not the same.

Alex started the conversation by saying the blocks were the same because they had flat sides—a good start. Steven brought up that they both had four corners. I asked him to show us what he meant, and he pointed to the four shorter edges of the pattern block as he counted, "1, 2, 3, 4" (see figure at left). Alex disagreed and said there were eight corners, pointing to each vertex as he counted (see figure at right).

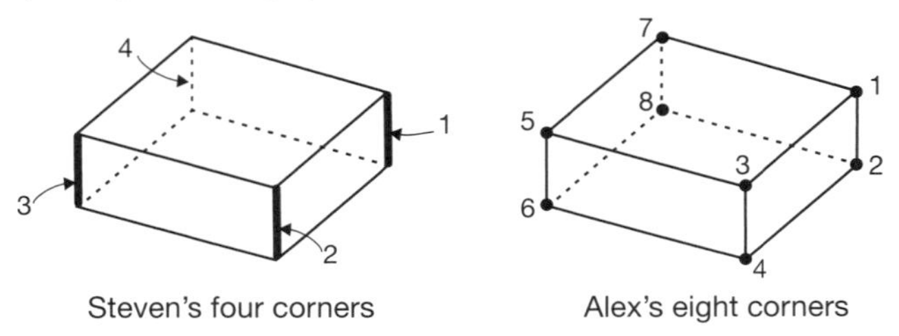

Steven's four corners   Alex's eight corners

All of a sudden, children were perking up and joining in. I felt I needed to make a decision about today's topic. Here was some real interest, but it was moving away from my plan for the rest of the activity. I kept thinking maybe I could just straighten it all out if I told them what a corner was. But, what is a corner? It isn't really a mathematical term at all. More important, I needed to find out what they each meant by the word. I resisted the temptation to tell them anything. There was something going on here. It felt like more than just a definition was at issue, but I couldn't pinpoint exactly what the mathematics was. I decided, for everyone's sake, to slow the conversation down. That would at least give me more time to decide how to proceed.

I showed what Steven was counting: the edge that went from the vertex of one face to the vertex of the other face. Trevor then illustrated what Alex was counting. Alex defended his count of eight, but Steven continued to disagree.

ALEX: There're two sharp corners here, not just one sharp corner.

STEVEN: No, it's one sharp corner. It's just a long one that goes to each side.

Steven's thinking was beginning to make sense to me; he was thinking about a corner as the edge that connects two vertices. So, what was a corner, then, and how many were there? We didn't agree—maybe four or maybe eight. Little did I know where my decision not to give a definition would lead us! I was working hard to figure out what mathematics, if any, we were talking about. I decided I had to pursue this.

I formed small groups and asked them to reexamine the two blocks we had been discussing (the square orange pattern block and one of the rectangular Geoblocks) to decide how many corners each block had. One group of three continued the disagreement. Holly decided on four corners, while her partners, Kathryn and Sarah May, arrived at eight. I told them I was interested in how they could look at the same blocks and yet count different numbers of corners. What did they think was going on? As we discussed this together, it was clear that Holly had the same interpretation as Steven, while Kathryn and Sarah May were, like Alex, thinking about corners as vertices.

It was still bothering me that I wasn't clear where this discussion had taken us. Were Kathryn and Sarah May dealing with a geometry concept or a counting problem? Could I have avoided this by introducing the terms *edge* and *vertex* and giving definitions? Would we then have avoided this contradictory discussion? And if that was the case, was avoiding this kind of confusion a worthy goal, or were the students getting something out of this confusion?

The more I thought about it, the more I felt satisfied with what had happened. We spent our last bit of time clarifying, as a whole class, what people were paying attention to when they were counting. We discussed whether we had one corner or two corners at the place where two faces met. Kathryn, Holly, and others shared their ideas with the group. In the end, many children were saying it could be two, but those two are connected, and it didn't matter because it depended on how you looked at it. Nikki then verbalized what you had to count to get four corners, and what you had to count to get eight.

155 160 165 170 175 180 185 190

So, even though they were using a nonmathematical term to describe aspects of the blocks, they had worked together as a whole class to understand what each group was trying to explain. I felt this issue was somewhat resolved because they realized that how they looked at the object made a difference in their count. That feels close to understanding why we need to have a common definition. A corner could be either thing, as long as we agree what we mean when we say corner.

I was also struck by the math they were doing. They had named significant features of 3-D shapes. Some were close to seeing edges, and others were noticing vertices, but all were paying attention to important aspects of the 3-D objects.

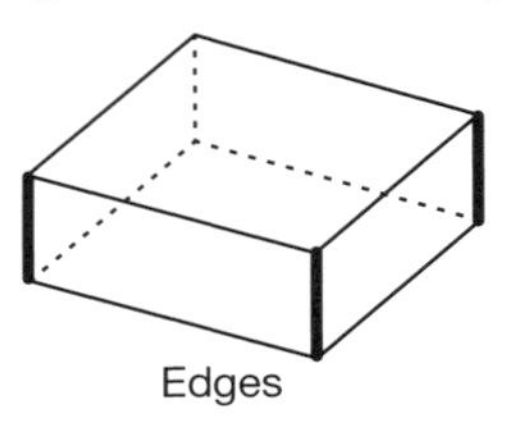

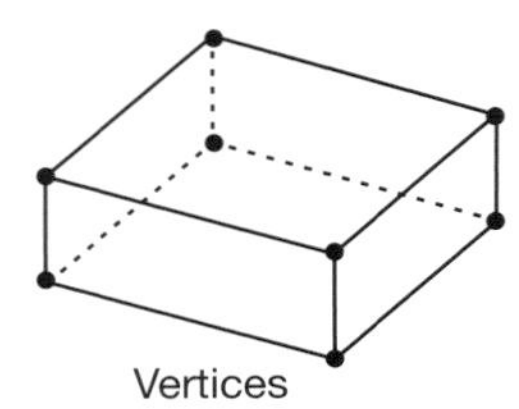

There is more to sort out about corners and edges and vertices as our geometry work continues, but this was a good beginning.

case 10

Different ways of seeing edges

Isabelle
GRADE 2, DECEMBER

Several weeks ago, my second-grade students spent some time exploring a set of three-dimensional wooden blocks. One activity required them to carefully examine the faces of the various blocks. On other occasions they have enjoyed building with the blocks during free choice time. I recently asked them to find and count the edges of the blocks. I showed them what I meant by the edge of a block, and then asked several students to restate the meaning of *edge* in their own words. We developed a common understanding that the edge is the place where two faces of the block come together. I chose to begin by looking at a cube and gave each pair an identical cube to examine. I asked students to use words, pictures, and numbers to show how they had counted the edges. 205 210

When we gathered on the rug to share our work, Fiona took the cube to show how she and her partner had counted the edges to arrive at an answer of 12.

FIONA: We counted all these sides. I counted by fours because there's four corners (edges) on the top [*pointing to the top of the cube*]. And then I counted this four [*pointing to the opposite end of the cube*]. And then I counted this four [*pointing to the four edges that join the side faces of the cube*]. 215

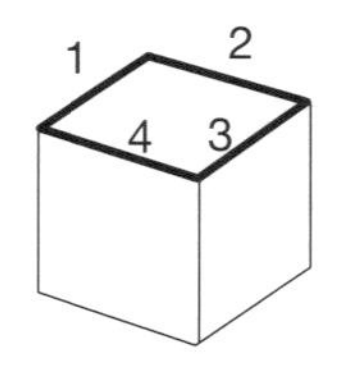

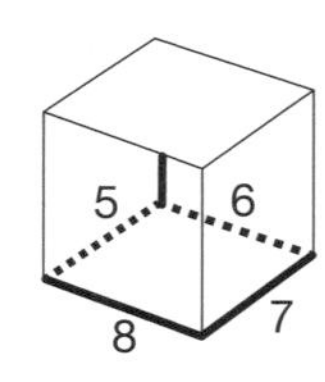

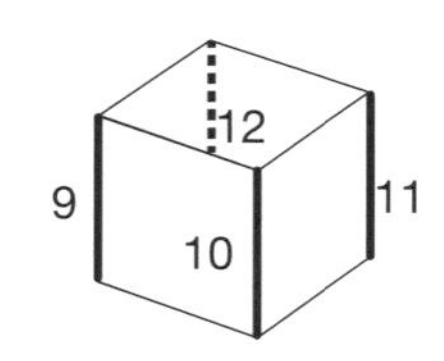

TRACY: We did something like that. We put the faces that we already counted down on the floor. 220

Tracy counted the four edges on top, turned the cube over and counted the four edges around the sides, then counted the four edges on what initially was the bottom.

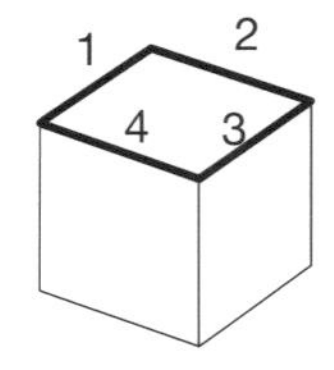

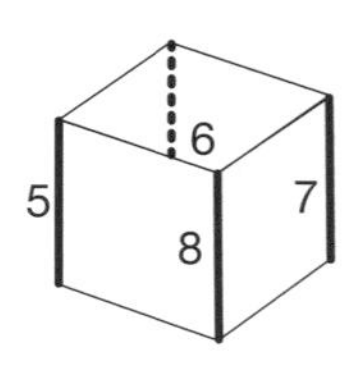

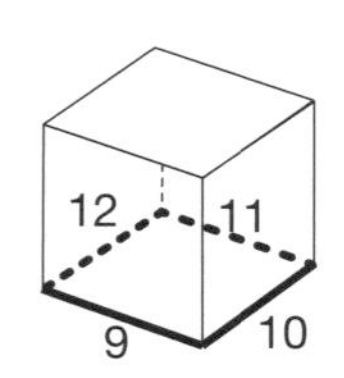

225

Seth and his partner had arrived at an answer of 24 edges. He shared how they had counted.

SETH: [*running his finger around the side of one face*] We counted 1, 2, 3, 4. Then we kept a finger on this one so we knew where we started. Then we went 1, 2, 3, 4 [*he runs his finger around the edges of adjacent faces, in sequence*]; 1, 2, 3, 4; 1, 2, 3, 4. Then we counted the top and the bottom. 230

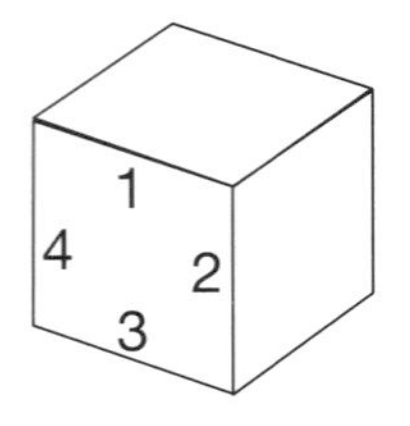

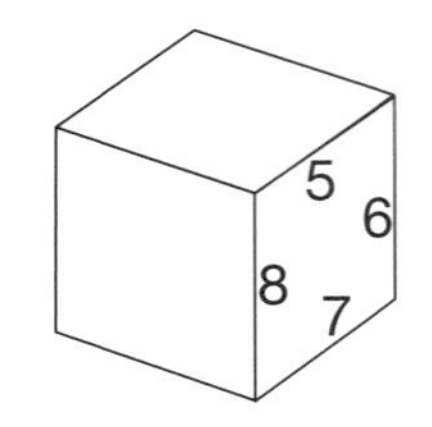

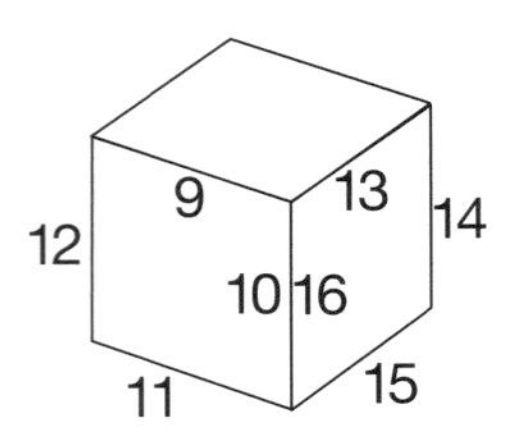

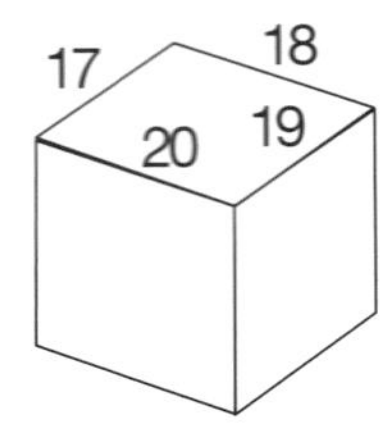

KARYN: I have something to say about that. If somebody counted this [*pointing to the edges of a face*], it would be 4. But then if they counted over here [*turning to an adjacent face*], they would count the same one again. This one and this one [*again pointing to two adjacent faces*] have the same point. 235

TEACHER: Can someone else say what Karyn is talking about?

TRACY: Each edge that you count, and then you go to the other edges and you count them, you count the same edge again.

FIONA: It seems that there's only three on each side, because you would be counting the same exact one again. 240

TEACHER: What do you mean there are only three on each side?

FIONA: These sides, you count three of them. Then if you go on to the fourth one and go on to the next one, you'll count the same exact side again.

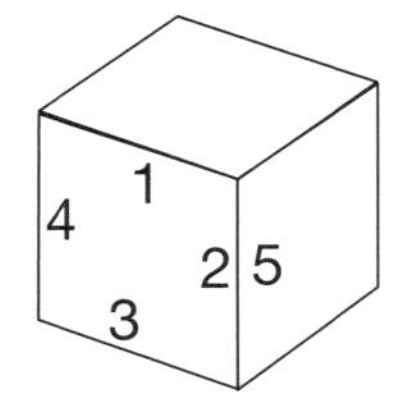

TEACHER: So what do you do instead? 245

FIONA: And you're just making more of them. You're making more sides if you take this four and count this one again, you're making more sides.

TEACHER: So how do you solve that problem?

FIONA: You only count three, and then you go on to the next side. [*She runs her finger around the edges.*] 250

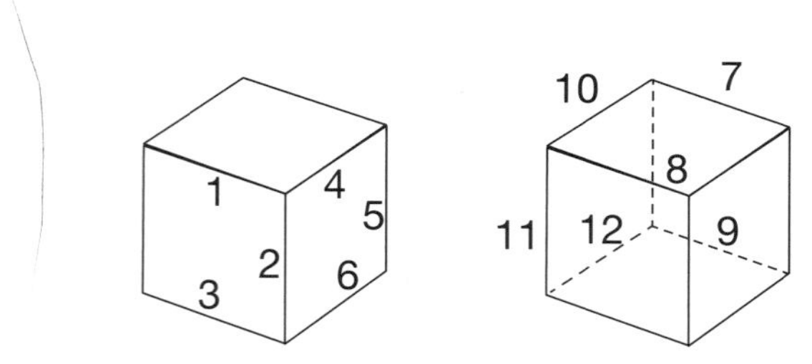

Fiona seemed to be noticing that the edge of a cube is the place where two faces of the square sides meet. She had developed a way of counting that allowed her to look at a cube and take the "shared edges" into account. Seth, on the other hand, was noticing four edges for each of the six squares. So he was counting four edges six times. 255

As I watched and listened to my students during this discussion, I thought they were struggling with counting systematically. I also saw that they are using terms like *edge*, *side*, and *point* in ways that are inconsistent and that they are not using the conventional meanings for those terms. Now I see that they are grappling with these ideas: what is an edge, and what happens when the sides of two squares come together to form the edge of a cube? These ideas 260 will be the focus of future lessons.

case 11

Taking a second look

Janine
GRADE 4, NOVEMBER

It was time to start a unit on geometry, but I realized my class had had little previous experience with three-dimensional geometric solids. To give them more experience, we spent a class session exploring how these objects were composed and talking about the differences between 2-D and 3-D objects. 265

We examined a set of twelve geometric solids. I drew their attention to the faces, edges, and vertices of each object. I asked the students to work with a partner to draw each one and to record the number of faces, edges, and vertices of each solid. They very excitedly got to work. Doing geometry somehow seemed really important to them, very grown-up! 270

As I walked around the room, I noticed several things. My fourth graders had a lot of difficulty drawing these 3-D shapes, even when they were copying from a sheet on which the same solids were already drawn in two dimensions (though they didn't have much trouble identifying each solid from the sheet). I also noticed that for some solids, they had a difficult time counting 275 the edges. They didn't have a system for counting and often counted the same edge twice.

I realized quite suddenly that what they were doing—what I had instructed them to do—was not really very useful. They were looking at the solids carefully, but I now realized I should have told them not only to count, but also to describe the faces. For example, students were writing about a cube, "It has 6 faces, 12 edges, and 8 vertices." What they really needed to notice was that all the faces were squares. So, as I continued to circulate, I asked pairs to also describe the shape of the faces and explain how the faces connected to make the solid.

To see if this additional instruction helped, I sat with a group that had just begun comparing three objects that they had identified as a cube, a square prism, and a rectangular prism.

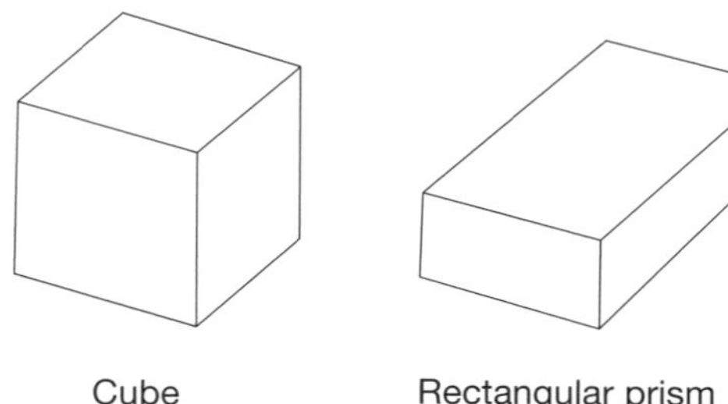

ANNA: [*picking up the rectangular prism*] This looks flatter than a cube. It looks more like an ice cube. The cube is taller.

EBONY: Right. The cube is taller, and the rectangular prism is more "shrinkish." It looks almost like half of a cube. In the cube, all the faces are square.

TEACHER: Can you talk about the square prism?

ANNA: In the square prism, the side faces are more . . . scrunched. The shape of the sides is more squished.

EBONY: The side, top, side, and bottom are all rectangles, and there are two, two-dimensional squares connecting them.

DARIA: Yeah. The square prism is a shape that has two squares with four rectangles connecting them. That makes it three-dimensional.

TEACHER: What makes it three-dimensional?

ANNA: Because if you take the top alone or one of the sides alone, they are two-dimensional, but if you put them all together, they are three-dimensional. They are, like, "deep."

EBONY: [*picking up the triangular prism*] This has two kinds of shapes. Rectangles—leaning—and triangles.

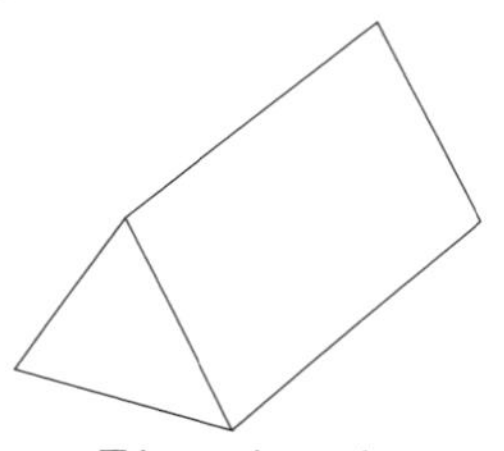

Triangular prism

TEACHER: Can you tell me more about it?

EBONY: Well, it has three rectangle faces connecting two triangle faces. It stands on one rectangle face, and the other two rectangle faces are leaning. I guess if you tipped it up, it could stand on a triangle face too, and it would be real tall! 305

I was happy to see them looking closely at the shape of the faces, and also happy with the way they were starting to look at the object as a whole. By simply counting the faces, they would not have accomplished this much. For some shapes, though, they had a hard time trying to explain what they saw, falling back on words like "shrinkish" and "scrunched" and "squished." They were also using their hands and their faces to help explain what they were seeing. There were lots of pauses and lots of think time. 310

EBONY: The triangular prism is a little like this one [*holding a square pyramid*]. This one is like Egypt! It has four triangle faces going up. It has edges going up, too. If you push the triangles together, it makes a pyramid like the ones in Egypt. The triangular prism has two rectangle shapes leaning together at the top. The pyramid has triangles that go up more. 315

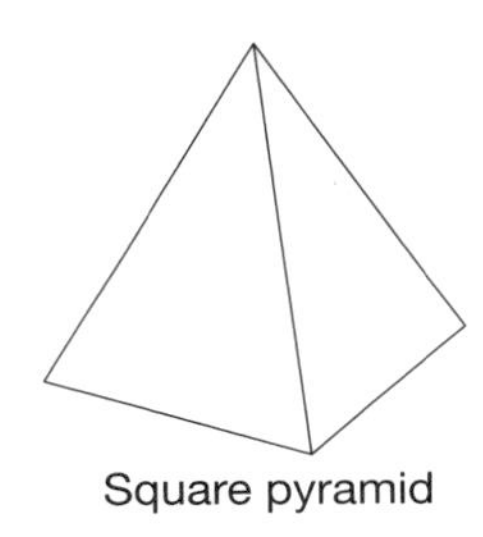

Square pyramid

TOMAS: What's the difference between the two? 320

JOE: One's bigger.

EBONY: In the pyramid, it looks like the side faces are slanted more. They are leaning more.

JOE: The pyramid is bigger and it has a pointy top. Pushing the triangles in makes it a pyramid. 325

ANNA: Yeah. The pyramid seems more folded . . . taller.

EBONY: The pyramid has one point at the top. That makes the pyramid more pointy.

JOE: Look! [*He spins the pyramid around with his hands.*] You can twirl the pyramid and it looks the same on all the sides, but you can't twirl all of the prisms and have them look the same. 330

Everyone thought that was pretty cool and took turns twirling the pyramid. Actually, I thought it was a pretty cool observation myself!

EBONY: I think . . . yeah, the square pyramid is standing on a square. I bet that's why its name is square pyramid.

I asked the group to think about a definition for a prism, because in reality, I wasn't sure of an exact definition myself. (The word always reminds me of playing with prisms as a kid, prisms from a candelabra that refracted light into tiny rainbows.) My fourth graders decided "a prism is like a cylinder that has more faces around it." As an example, they described a hexagonal prism. They said it looks just like a cylinder, except it has six faces around it and a cylinder has only one face around it. (They didn't talk about the bases—circles on the cylinder and hexagons on the hexagonal prism.) 335 340

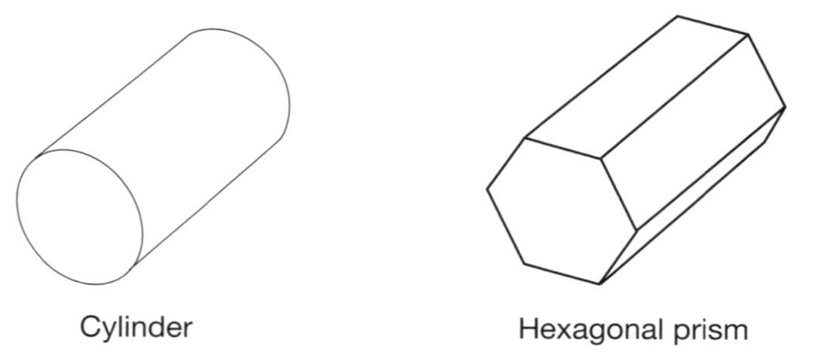

When I looked up prism in a dictionary, I found that it was a "solid with bases or ends that are parallel, congruent polygons and sides that are parallelograms."

I had learned a lot from listening to my students try to make sense out of the geometric solids, especially observing their struggles to find words for what they saw. I was really impressed with the twirling comment and actually went back myself to see which of the other prisms could be "twirled," and what that said about them. 345

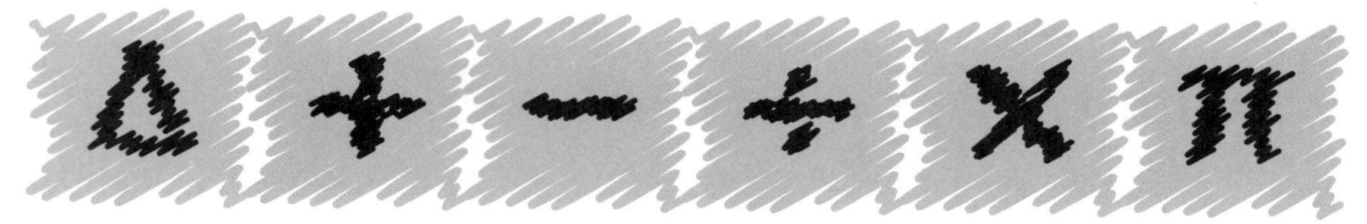

CHAPTER

3

Making sense of angles

One attribute of shape that children notice is angle. They see that some corners are more "pointy" than others, and often know that some corners are right angles. However, understanding just what an angle is, is quite complicated. Before you read the cases in this chapter, jot down your own answer to this question: "What is an angle?"

How do children make sense of angle? You'll explore some fifth-grade thoughts as Nadia's students respond to a classmate's question, "What are angles?" Dolores's third graders, Sandra's seventh graders, and Jordan's fifth graders are all concerned with angle size and how to measure angles. The work of Lucy's third graders leads us to questions about the relationships among turning, rotation, and angle. As you read the cases, identify what children bring to their study of angles.

- What everyday experiences do they draw on to explore this concept?
- What confusions do they need to work through?
- What are the connections between what an angle is and how we measure it?

After you have read the cases and considered these questions, reread your own definition and, if you wish, add an addendum.

case 12

Stumbling upon names for angles

Dolores

GRADE 3, JANUARY

This year my third graders have done several geometry activities. Our most recent work involved triangles. In this work, children noticed that triangles have three corners or, as some said, three "points." (I had used the word angle, but that more technical term seemed unimportant or unnecessary to my students.) They had also noticed shapes that they said were "too pointy to be triangles." Their greatest comfort level was with equilateral triangles pointing upward.

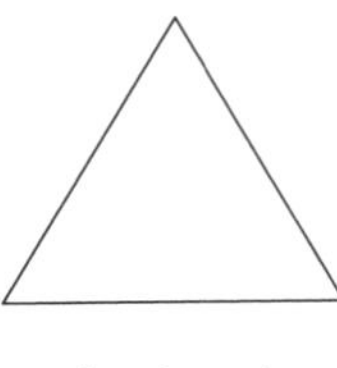

A triangle

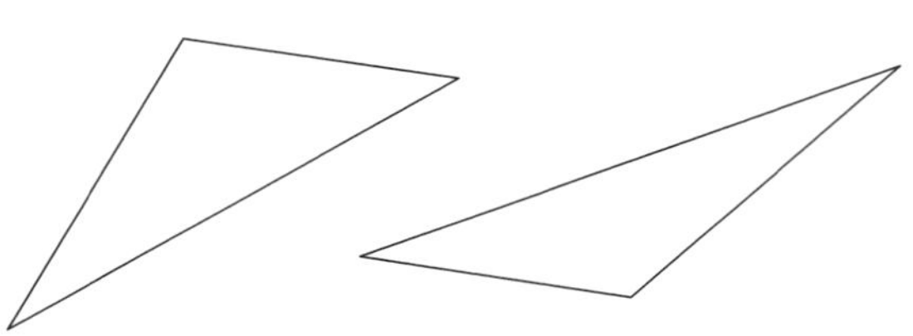

"too pointy to be triangles"

Some days later, and quite by chance, someone was looking up *angel* in the dictionary and found *angle*! Word spread quickly through the room that the dictionary had pictures illustrating three different kinds of angles. Children crowded around the open book. Soon others were searching for the word *angle* in their own dictionaries for a closer look.

The pictures showed a right angle, an obtuse angle, and an acute angle. I knew that I had to capitalize on this spontaneous interest and energy. We began by looking around the classroom to see if we could find angles of these kinds. Someone soon discovered that taking an ordinary sheet of paper and holding a corner of it up to an angle would help us decide if the angle was right, obtuse, or acute. This "corner-of-a-sheet-of-paper" measuring tool became popular. Children formed little groups and moved about the room, finding angles, sizing them up with their right-angle paper corner, and then consulting the dictionary again for the right term to describe what they had found.

A week later I decided to see what the children had learned from their spontaneous investigation of angles. On the backs of their spelling tests, I asked them to take a few minutes to explain something they knew about angles. I was surprised by what I found. I've included some samples to show what the children brought away from the experience.

Chad is generally a talented mathematics thinker. I was interested in his choice of the words *long* and *short* (see fig. 3.1). Is he noticing the length of the sides that form the angle? Clearly the sharpened part of a pencil tip is a small angle, but the sides that form the angle are also short.

Comparing a pencil tip to arm lengths is somewhat unclear. I tried putting my arms together and found I can actually make many different angle sizes. I'll need to speak to him, to clarify what he is paying attention to, the sides or the actual angle. 30

Cindy (see fig. 3.2) seems to have a good sense of the variety of directions in which angles can be oriented. I can't tell from what she has written if she knows anything about the variety of sizes. Her use of the word *slanted* will be the opening for a whole-group discussion. I often take a phrase from one child for discussion with the whole class to find out if others are thinking in a similar way. 35

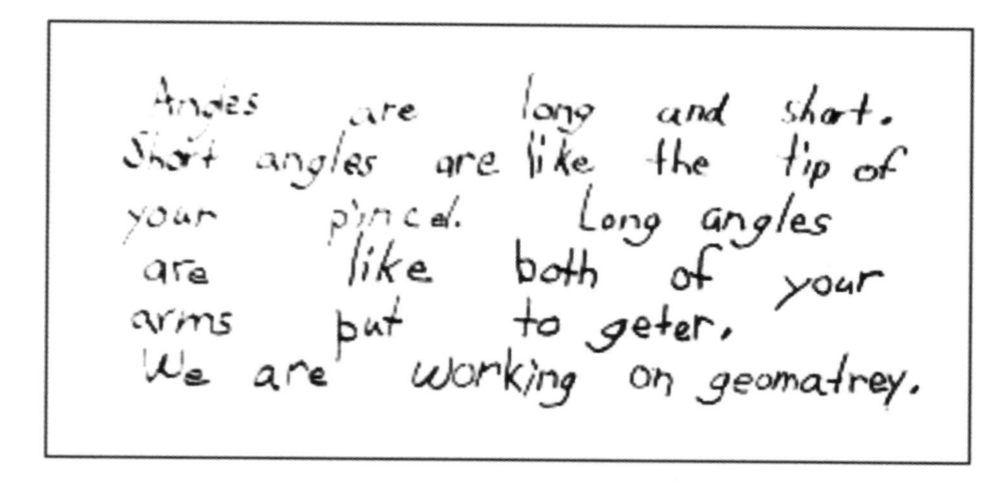
Angles are long and short.
Short angles are like the tip of
your pincel. Long angles
are like both of your
arms put to geter.
We are working on geomatrey.

Fig. 3.1. Chad's writing about angles

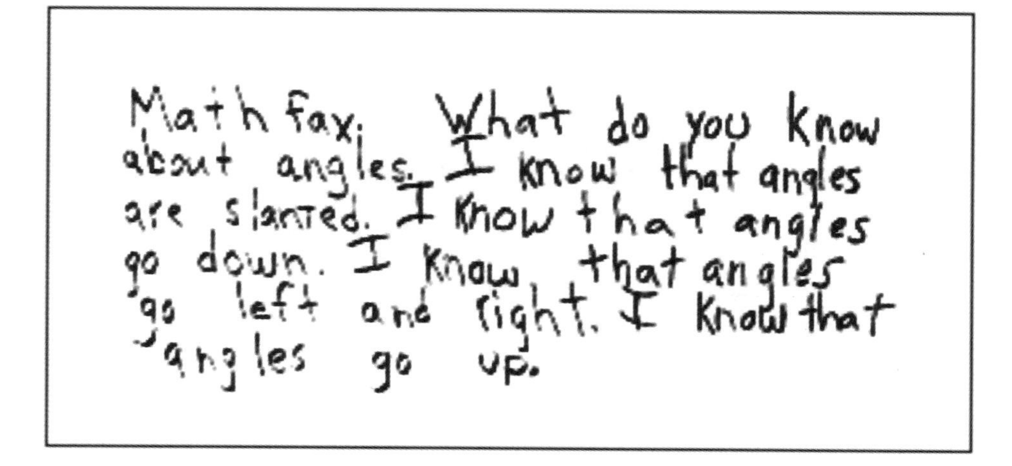
Math fax. What do you know
about angles. I know that angles
are slanted. I know that angles
go down. I know that angles
go left and right. I know that
angles go up.

Fig. 3.2. Cindy's writing about angles

Nancy's comments (see fig. 3.3) remind me that an ordinary pencil is never far from a child at any part of the school day. When the search for angles broke out, the pencil point was one of the most often mentioned angles. Nancy tends to pay close attention to detail, so I'm not surprised she connected acute and little. Not so clear is what she understands about right angles and obtuse angles. It would be easy for me to think she's really onto something with her statement "Angles can be big." She could be referring to the size of the angle or to the length of the sides, as Chad did. I can see how these ideas would provoke a great discussion.

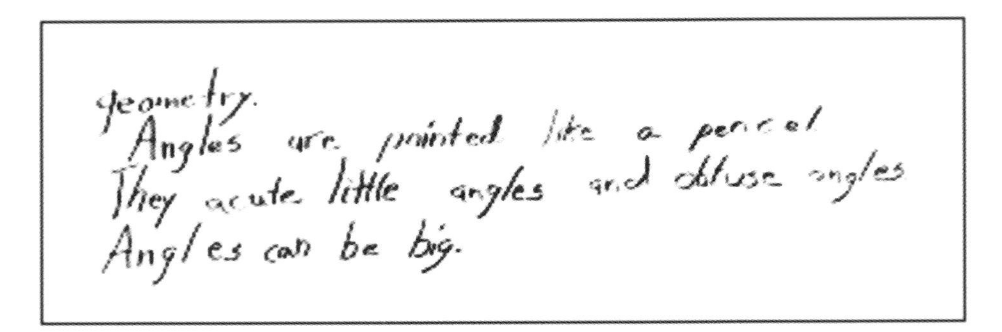

geometry.
Angles are pointed like a pencil
They acute little angles and obtuse angles
Angles can be big.

Fig. 3.3. Nancy's writing about angles

In math we're doing
shapes. Acut angels are
sharp Right angels are 90°
obtuse angels are above 90°

Fig. 3.4. Crissy's writing about angles

Crissy is saying things that really seem to make sense (see fig. 3.4). I just don't know for sure what the term *90 degrees* means to her. She was one of the first students to use the idea of the square corner on sheets of paper to compare angle size. My feeling is that she has some good foundation knowledge.

Chelsea really grabbed onto the idea of marking right angles with little squares (see fig. 3.5). Her sketch of a rectangle by itself seems to show great understanding. Once she goes further with her triangle illustrations, though, I'm less sure of what she knows. If she used the corner of a sheet of paper on her diagrams, would she see that the marked angles are not really right angles? Also, can a triangle have two right angles? What a wonderful question to pose to the group. I wonder what they'd say if I asked them that.

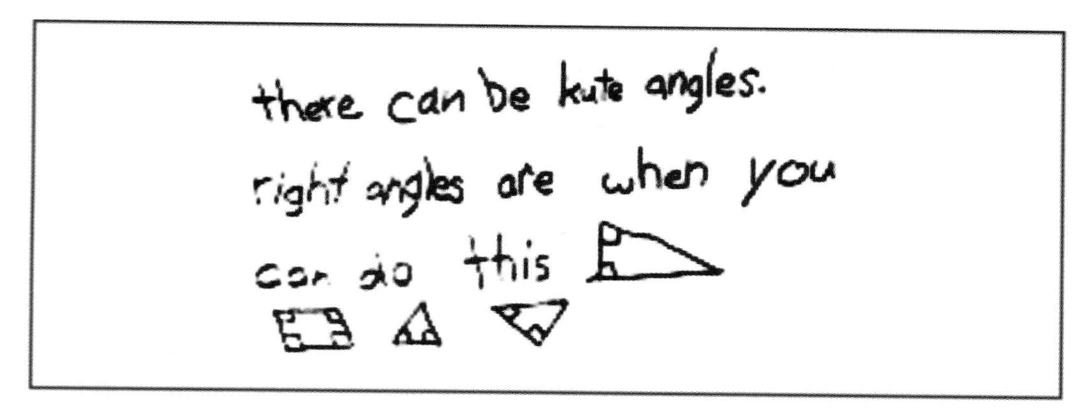

Fig. 3.5. Chelsea's writing about angles

As their writing makes clear, my third graders are beginning to notice angles. Naturally they vary in their degree of interest and in their understanding of angle sizes. I will need to decide if I should introduce tools to measure angles. This was just a start. For most of the children, it is a new idea. As I think about my third graders, I am remembering a conversation from a class I took with other teachers about geometry. The whole idea of defining an angle was hotly debated. What part is the actual angle? Is the angle the space between two lines that come to a point? Is the angle the two lines that meet at a point? Is the angle the lines and the space between? This is a complex idea for third graders and for teachers as well.

Since this experience, I have noticed that the world seems to be full of references to angles. Newscasters go to another reporter for "another angle on the story." Plans for new public buildings are "shown from different angles." Weather reports include references to the "angle of the sun." I wonder if the same thing happens for the children. Will all this talk of angles suddenly start to pop out at them? I'll be on the alert for places to investigate angles as the year goes along.

case 13

What are angles?

Nadia
GRADE 5, FEBRUARY

For a week now, we have been exploring 2-D shapes, finding and defining their features. Toward the end of our last class, Martha and Nicole were in disagreement over how many angles there are in a triangle. Martha claimed, "There are two angles in a triangle," whereas Nicole insisted, "No, there are three." Class ended on a question from Sophia, prompted by this disagreement: "What are angles?"

I've seen children confused when they encounter angles within the context of a closed figure. Sometimes children learn a symbol to represent the idea of an angle before they understand when and where we typically use such symbols. Consequently, children can easily identify these symbols as representing angles:

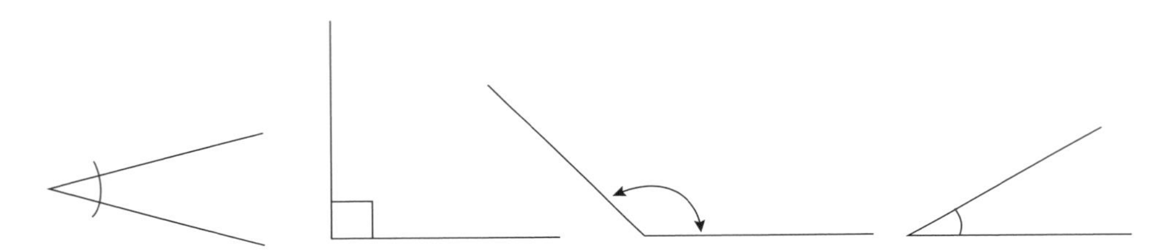

However, using these symbols in a larger context is harder, and my students don't always recognize the angles in a 2-D figure.

What was Martha thinking about when she argued that a triangle has two angles? Is it possible that she only considered the two "bottom" angles? These angles would resemble most closely the symbolic angle representations she had seen; the "top" angle might not look the same to her. I wanted to find out more, so today I started the class with Sophia's question, "What are angles?"

The children were enthusiastic and eager to share their ideas. Their responses varied in style and detail, but they all reflected the following ideas:

- Angles have two sides.
- Two sides form an angle.
- When two sides turn, they make an angle.
- "Degrees" is an angle.
- They are inside a corner.
- They are inside a shape.

After we had listed this last point, Laurel spoke up to argue, "I don't think that all angles are inside a shape." Laurel's objection brought an odd quiet to the class. The children were listening and waiting to see what was going to happen next. It really took us all by surprise. It was clear that the children heard Laurel and were thinking about what she had said. Were they trying to make sense of it? I waited a few seconds to see if anyone would say something. Instead, they all looked at me. They wanted to hear what I thought of Laurel's comment. I looked back at them, and then urged Laurel to explain further. 100

"Can I come up and show you?" she asked. Laurel had all our attention as she went to the chalkboard and drew this figure: 105

Pointing to the marked angle, she explained, "This is outside the shape, and there is an angle in there."

There was silence in the class, a good silence. The children were thinking and taking it all in. Kara, who doesn't like to be wrong, was surprised and reacted first. "But in a triangle," said Kara, "all angles are inside." 110

Laurel listened to her without responding. Was Kara trying to make sense of Laurel's statement? Was she trying to understand how this information fit with her prior knowledge? Was Laurel trying to find the "outside" angles in a triangle? While Laurel's shape was like a triangle with a dent in the base, it actually formed a quadrilateral. I wondered if the children had ever considered non-convex shapes before. I wondered if the children would have an easier time identifying this "outside" angle than they had seeing the "inside" angles in a triangle. I wanted them to see that Laurel's shape was a real geometric shape, a four-sided polygon. After all, it had line segments for sides and was a closed figure. I wondered if Kara saw Laurel's shape as a valid shape. 115

TEACHER: How is a triangle different from and similar to Laurel's shape? 120

KARA: The triangle has all sides the same, and it is all closed.

LAUREL: Are there angles outside a triangle?

KARA: No, the angles are inside a triangle, and all sides in a triangle are equal.

Kara was disturbed that her beliefs were being challenged. She seemed to be thinking only of an equilateral triangle and did not know what to do with this new information. Laurel, however, was in a different place. She was willing to consider new possibilities. She was comfortable with the idea of outside angles in her figure and was on her way to making sense of the outside angles of a triangle. She sure knew how to take a challenge! At this point, the earlier disagreement resurfaced. 125

NICOLE: But there are three angles in a triangle! 130

MARTHA: No, there are two.

I asked Martha to show us the angles in a triangle, and she drew these figures on the board:

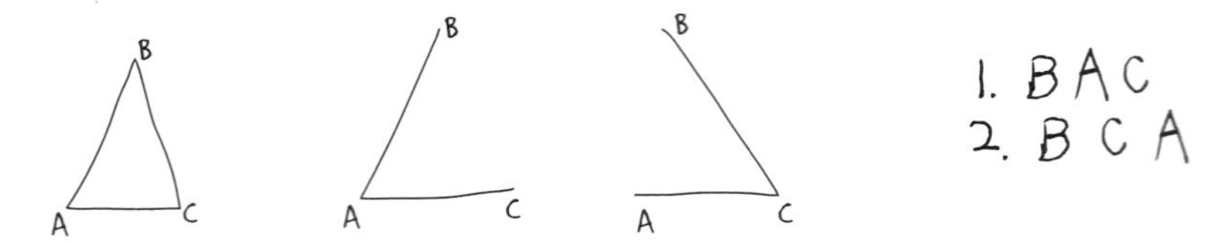

First she drew triangle ABC, explaining that the two sides of the triangle that meet at the top are the angles. Then she cut the triangle apart, showing the angles BAC and ACB. Now I knew that she has never recognized ABC as an angle. 135

Alana had her own ideas about angles that she asked to show us. She came up and drew three lines, explaining, "These are all angles. All these lines are at an angle."

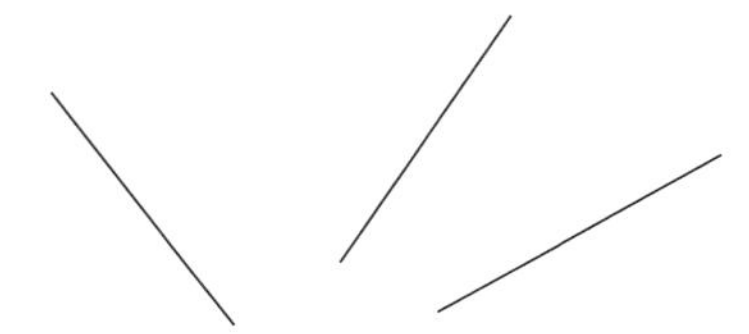

While Martha and Alana brought up significant ideas about angle, their comments also helped me to see what might be missing. I see now that the children have many different ways of thinking about angles, and some of their ideas seem more related to geometry than others. They were struggling to fit their understanding of angles to these figures. 140

The symbolic representation for angle $\angle$ doesn't seem to be present in these figures at all. Thinking outside the box of standard examples of shapes and angles may be hard for them because they haven't had many experiences with shapes. I need to increase their exposure to angles in many different contexts to help them expand their working definition. For example, exposure to different contexts might help the students understand angles as an intrinsic element of shapes rather than as isolated fixtures with no meaning nor significance in the shape of the figure. I want to see what they will make of the angles in these figures: 145 150

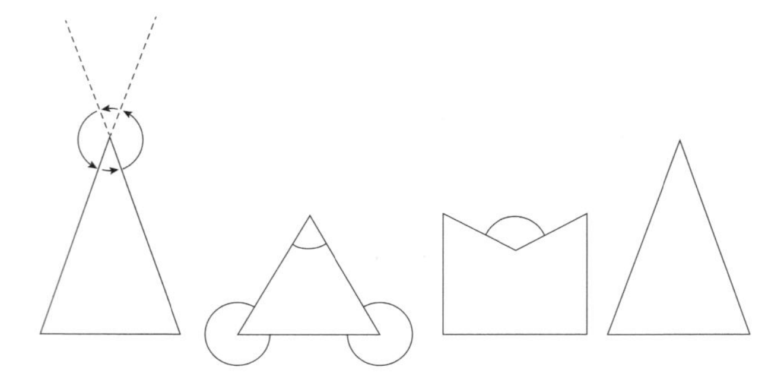

Alana's interpretation of lines "at an angle" is another common view that needs to be expanded. In such situations, I need to ask the students, "What makes the angles of these lines?" or "At an angle with what?" I know that "at an angle" means the lines make an angle with the horizontal line of the ground. For example, if you hold a ruler at 60 degrees, you mean that the ruler makes a 60 degrees angle with the horizontal. But I ask myself, what experiences and activities will help my students make sense of this idea? 155

Culture, language, and self-understanding influence what we know about the world. In spite of my awareness of what my students still need to work on, I was impressed by how hard they were thinking about angles. Their comments helped me see just how complicated it is to define angles. When I was a fifth grader, I wasn't asked to think about questions like "What is an angle?" I learned the rules. I was told where to find angles in figures and how to measure them. The rules made the definition for me. I never had the chance to question or think for myself about what angles meant and what role they played in the shape of the figures. When I was told in my high school years that there were angles outside of figures, I was shocked, and it took me a long time to understand this idea. 160 165

My students were struggling to understand and define what angles are, but they were on the right track. They were encountering different contexts for angles, starting to see them both inside and outside of figures, seeing lines "at an angle," and trying to determine how many angles are in a given figure. As these lessons continue, I want students to see how angles determine the shape of a figure, to recognize where the angles are in a figure, and to explore how angles are measured. I feel very excited to see how invested they are in discussing angles. That feels terrific! 170

case 14

Angles and turning

Lucy
GRADE 3, DECEMBER

Since the start of the school year, my class has been using a shape-making computer application. They've learned various commands, and have discovered how to make the computer draw squares. They've learned that turning the cursor 90 degrees right or left will create square corners. They've also explored the amount of turning necessary to create angles of different sizes. For years, I have wondered how third graders understand angles and turns. Now I wanted to find out more about what they were thinking about angles. 175 180

We began with an assignment to cut pictures from magazines that showed angles of different sizes. I asked them to decide whether each picture showed an angle of 90 degrees, less than 90 degrees, or greater than 90 degrees. Finally, each group was to create three posters, showing how they had sorted their pictures. Because they had noticed that each picture actually included many angles, I asked them to mark the angles they were sorting by. 185

I was surprised by the finished work. When I looked at the posters, I had many questions. Often I was unable to tell how the pictures had been sorted. To my eyes, several posters showed angles of all three sizes. One poster for More than 90 degrees had a picture of a skater bent over at the waist holding her ankles. The group had marked the angle formed between the upper part of her body and her legs. It looked like 30 degrees to me. Next to the photograph of the skater were pictures of a breast cancer ribbon and a star from the Hollywood Walk of Fame: 190

All three of these were on a poster headed "More than 90 degrees." Another poster labeled "Less than 90 degrees" had many photos of the angle formed between peoples' legs. But also on that poster was the angle made by the roof of a house that appeared to be about 150 degrees. I wondered what the children were really seeing and thinking as they sorted the pictures. I decided to begin the next class by displaying the posters and asking the class what they thought. 195

ADAM: [*pointing to the picture of skater*] I don't think that is more than 90 degrees. That is a lot less than 90 degrees [*pointing to the space between her midriff and thigh*]. 200

TEACHER: Any comments?

SARAH: I think it is more than 90 degrees because she's bent so far over. Let me show you. [*She stands up.*] It would be 90 degrees if she was bent this far. [*She bends at the waist to approximate a 90 degrees angle.*] But she has to turn farther. She is bent more than that. So it's more than 90 degrees, like maybe 120 degrees. [*By now she is bent way over, just like the skater.*] 205

Sarah was paying attention to the action that created the turn. This surprised me because I had initially had the same reaction Adam did to that angle. The angle made by the skater's body was quite small, much less than 90 degrees. But what Sarah did, showing how the turn created the angle, made a lot of sense to me! 210

ADAM: But that space isn't even 90 degrees. [*He points again to the space between her thigh and midriff.*] If it were more than 90 degrees, then she'd only be bent a little. [*He shows us by bending a bit from the waist.*]

RON: Maybe it can be both less than 90 degrees and more than 90 degrees.

215

In a sense, Ron was correct. It depends on which angle we are looking at.

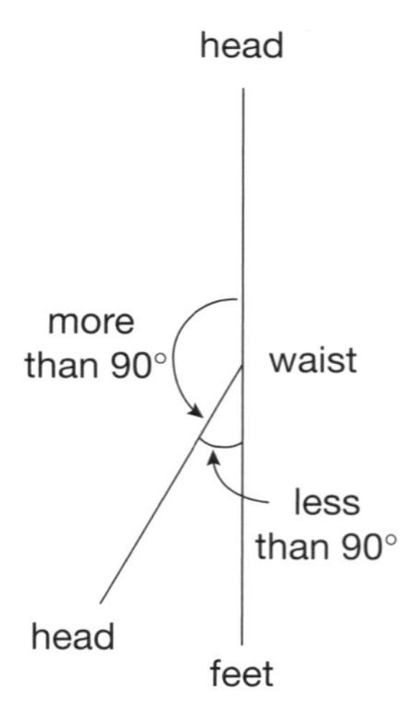

As the skater bends, the turn is more than 90 degrees. But after the action is over, the angle between her upper body and her legs—the angle resulting from that amount of bending (the angle actually shown in the picture)—is less than 90 degrees. I was surprised by the discussion. 220 It made me realize that some students' work was confusing to me because they were paying attention to different things as they sorted the pictures. They were noticing the movement implied by the picture and not the angle that resulted from that movement.

This made me take another look at the posters and the photographs they contained. Some photos had indeed captured motion; at least, there were implied movements by the people and 225 animals we saw in the photographs. Did the children think of these pictures differently from those of static objects? If movement was implied in the picture, did that make the children think about the amount of turn that created the final angle? In my mind, there is definitely movement in the creation of angles.

There was a lot to think about. The class had done no measuring of angles, except in the geometry app. There, they looked at angles in terms of how far to turn the turtle to create the angle. We had never discussed the connections between the turn and the angle that is drawn. No wonder they were so inconsistent. Sometimes they were thinking about a turn, or a change in direction, and sometimes they were thinking about the angle they see after the turn is made. 230

In hindsight, I wonder what would have happened if I had asked them to pay attention to the amount of turn. All year we have talked about the cursor "turning" so many degrees. Do they think of turns and angles as the same? When they make polygons by moving the cursor, are they thinking of the interior angle, the exterior angle, or both? I'm reminded of Ron's comment about an angle being both less than and more than 90 degrees at the same time. What are children paying attention to: the movement or the result? Should it be one or the other? Such a lot of fascinating questions. 235 240

case 15

Pattern block angles

Sandra

GRADE 7, NOVEMBER

There are so many parts to understanding angles! Is the angle the lines that intersect? Or are the lines just the boundaries for identifying the angle? Is the angle the space between the lines that intersect? A straight line represents a turn of 180 degrees, but is it an angle? What are degrees, anyway? My students frequently ask questions like these when we are studying angles.

As I was preparing for a unit about angles and polygons, I wondered how students recognize different shapes and how they develop a sense of measurement (both visually and with tools) to classify shapes. How do they adapt their thinking to include measuring line segments and measuring angles? What is it about angles that causes discomfort? I decided that based on the results from a pretest I had given in September and on my observations of students, I would need to give the students multiple opportunities to explore angles and polygons.

One activity we did together was both simple and fascinating to the students. I had them all cut out a paper triangle—any shape triangle they wanted. We all checked around the room to see how different our triangles were from each other's. Then I told them to cut their triangle into three pieces so that the angles could be separated. Next I had them fit the pieces together by placing each vertex at the same point, with the sides of the angles adjacent.

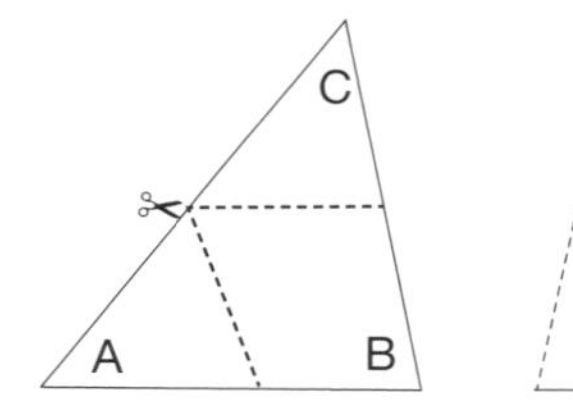

There were ohs and ahs as they looked around to see that every triangle could be rearranged to fit along a straight line. We concluded that if a line is 180 degrees, so must be the three angles of a triangle.

Another exploration was to find the measure of each angle in each pattern block.

 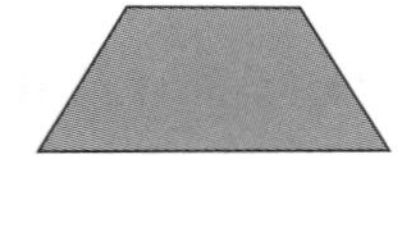 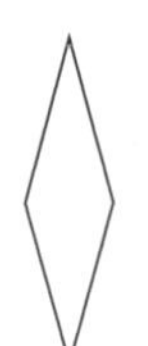 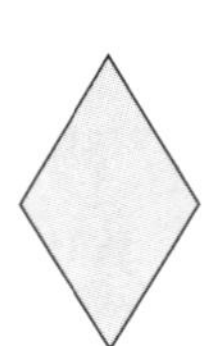 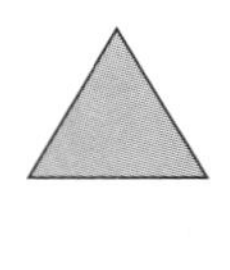 

I noticed that Casey, who had missed the first day of this activity, was struggling to make sense of what the others in his group were saying. He generally takes longer than others to understand new ideas. Because the others had tried to help him but were ready to give up, I decided to step in and serve as his partner for a time, letting the others go on. 265

CASEY: I don't get where they see the 120 degrees angle in these pattern blocks. They all look the same to me. The only angle I know is 90 degrees, and these all match my 90 degrees. 270

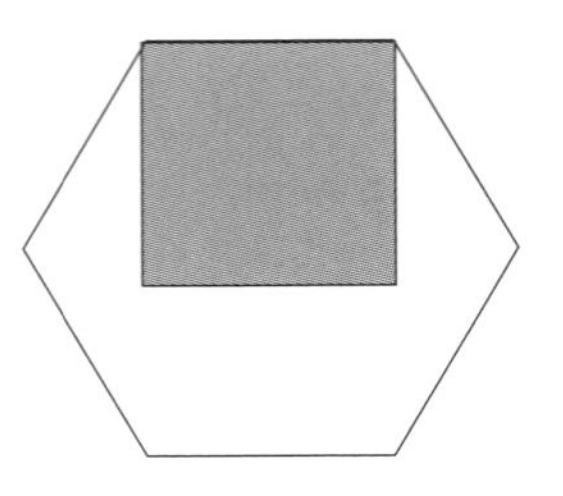

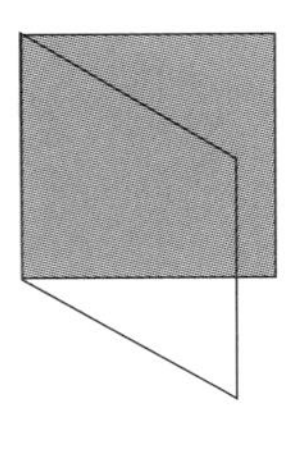

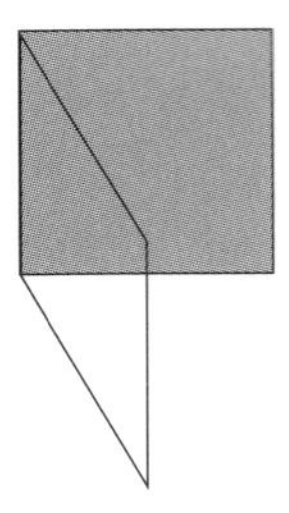

He placed the square on the hexagon so that the edge of the square matched one edge of the hexagon, and then proceeded to match the square's edge to one edge of every other block. Casey didn't seem to notice how different the angles were in the blocks. I decided to take him back to the idea of angles as turns, something we had done previously. 275

TEACHER: Let's review what we know about angles before we go any further with the blocks. Show me a 90 degrees angle.

As we had done in a previous lesson, Casey stood up with his left arm extended out to the side and his right arm sticking straight out in front. 280

TEACHER: Keep your hands glued to those spots. Turn 90 degrees left . . . turn 90 degrees right. Good. You really know those angles!

Then I repeated the movements. Casey agreed that I had also shown 90 degrees correctly.

TEACHER: Clearly, I'm taller than you and my arms are longer, so how can you say that I made a 90-degree angle? Is it the same as yours? 285

CASEY: Of course, it's the same! [*He looks at me as if to say this question is a no-brainer.*]

TEACHER: So how could we relate the 90 degrees we just made to the 90 degrees you found in the orange square? 290

CASEY: Let's see [*bending over the table with the blocks*], if my head was at this corner of the block, then my arms would reach down these two sides. I'd have to be really small to fit on this block.

TEACHER: [*running my finger down the edge of the block*] Is this the 90 degrees angle?

CASEY: Oh! That's not an angle. It's just one of the sides. Let me see. 295

I suggested that he trace the hexagon on a separate sheet of paper and look at one of the corners.

CASEY: Hey! The square didn't fill that corner.

TEACHER: Let's look at the hints given on the worksheet.

CASEY: [*reading from the sheet*] "Each angle of the square is 90 degrees. An acute angle of the white rhombus is 30 degrees." 300

When he tried to fill the remaining space with the acute angle of the white rhombus, it fit nicely.

CASEY: The square was 90 degrees and this piece [the white rhombus] is 30 degrees, so this must be what the others meant by 120 degrees. 305

I wondered if he was OK with adding the angle measures or if he was happy just landing on a number that the others had been talking about. I decided to let him try a few more tracings on his own. As I watched from a little distance away, Casey traced the white rhombus and tried to fill in its obtuse angle with several of the acute angles. It was important for him to place two of the blocks along the edges of the angle and then fill in the rest of the space with three 310 more blocks. Then an interesting thing happened. He removed the last three blocks and slid the square into the same place. He seemed pleased with himself. He then tried to replace the square with a blue rhombus but it didn't quite fit. So he tried the triangle and then the white rhombus to see if it would fit the space.

I didn't have a chance to revisit Casey during that class period, but the next day I noticed 315 that Casey was fully included in his group's activity. I am pleased that he figured out these pattern-block angles. I still have doubts about what Casey knows about angle measure. Although he was able to manipulate the pattern blocks to name some angle measures, this clearly was a first experience with angle size for him. He'll need many other opportunities to examine angles and to see how they are measured as the year goes on. 320

case 16

Measuring angles

Jordan

GRADE 5, MAY

Although my fifth-grade class had not formally studied angles, our various explorations in other areas had led us into much discussion about them. As the year went on, we came to "know" a few things about angles, including the following:

- A right angle is 90 degrees.
- Angles are measured in something called degrees.
- A full circle is 360 degrees.
- A straight line is 180 degrees.
- Some angles are wide and some are narrow.

Some of our information came from the skaters in the class who were more than willing to demonstrate their turns on a skateboard. Once the students' awareness of angles was heightened, they began to notice angles, comment on them, and want to investigate them more.

While studying the historical importance of a local canal that used an inclined plane to raise and lower boats from the canal to the river, we read that the angle of the inclined plane was 30 degrees. The students wanted to know what a 30 degrees angle looked like so they could get a better picture in their minds about the slant of the inclined plane. One boy explained that the corner of a square (or the corner of a sheet of paper, or a desk, or a book) can be used to judge a 90 degrees angle, so it's easy to figure whether an angle is less than or more than 90 degrees. His statement made sense because up to this point, we had just been eyeballing an angle to estimate its measurement, using the right angle as a benchmark.

Over the next few days, the students worked at an assignment to design pictures that included at least three angles: a right angle, an angle less than 90 degrees, and an angle more than 90 degrees. They estimated the measurement of the angles based on their knowledge of the right angle.

In Ellen's drawing (see fig. 3.6), the top of the picture frame (angle A) and the chair the boy is sitting on (angle G) are her benchmarks for 90 degrees. She estimated angle B at the base of the easel as 100 degrees, stating that it "looked like it was open about 10 more degrees than the 90 degrees angle." Ellen decided that angle C looked like a 45-degree angle because it seemed to be about half of the 90-degree angle, and she knew her estimate of 180 degrees for the back of the chair (angle F) was correct because it was a straight line.

Fig. 3.6. Ellen's drawing shows a variety of angle sizes.

I decided to introduce the protractor so they could measure with more accuracy. I wanted them to experiment with the protractor and the pictures they had drawn without any more directions from me. 355

TEACHER: We have been looking at angles and estimating their sizes. Now I am going to give each of you a special tool, called a *protractor* that can help you get an accurate measurement of the angles that you have drawn. You will have time to try it out to see how it works and to share ideas with your group. As you figure out how to use it, think about what you are paying attention to. 360

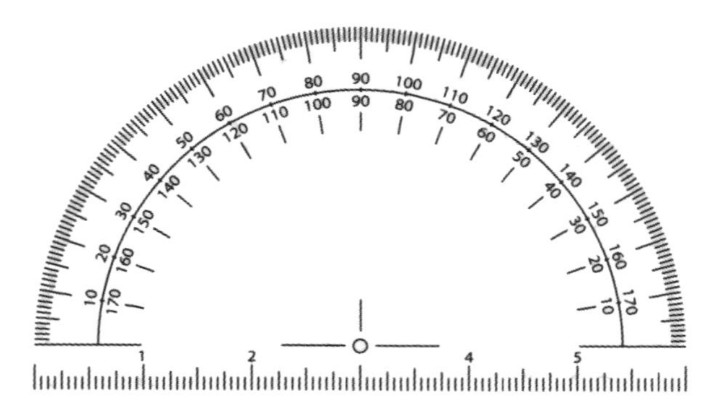

As I checked in with groups, I noticed that many students focused on the right angle first because they knew it was 90 degrees. That is, they knew they had to move the protractor in such a way that it read 90 degrees. It was the first time that most of the students had used a protractor, so manipulating it was a challenge. The fact that they routinely work in groups of four and share ideas with each other really helped. When someone figured out something about the tool, like how to place it on the paper relative to what they were measuring, they shared that discovery with the rest of their group.

After a period in which students worked individually and shared their frustrations, questions, and discoveries within their groups, I called them together to share ideas as a whole class. We compiled a list of statements about the protractors and the way to use them:

- One side feels smooth; the other side is bumpy.
- If you hold it a certain way, the numbers look backward; you have to hold it with the smooth side up so you can read the numbers.
- You have to pay attention to the numbers.
- There are two sets of numbers—low to high and high to low.
- The numbers in both sets are the same, just going the opposite way.
- The straight part at the bottom looks like a regular ruler.
- The rounded part of the protractor is where you get your angle measured, but you need the bottom part to line up with one of your angle lines.
- You have to line up the hole/dot on the bottom part with the point (vertex) of the angle on the paper.
- Make sure the dot is lined up with the corner of the paper.
- My protractor doesn't have a hole; it has an arrow, so just put the arrow at the place where the two lines come together.
- Make sure the straight part is resting on one of the lines, while the hole or arrow is at the point (vertex) on the paper.
- The numbers go from 180 to 0 and 0 to 180.
- The lines are important. If you look at where it says 100, it goes up by 1: 101, 102, 103. But if you go down from 100, it goes 99, 98, 97 . . .
- Each line is 1. The line that sticks out a little more is the halfway mark between the two numbers . . . like when we measured in centimeters.

During our discussion, one of the students wondered aloud how you know which number to use when it "lands" on 120. "Would you call it 60 degrees or 120 degrees?" These are some of the responses:

- "Picture 90 degrees in your head, then think, is it more or less? If it is less, then go with the lower number. If it is bigger, go with the bigger number."
- "Choose the number that is closest to your prediction."

- "Say you had a very narrow angle [*index finger and thumb are pinched almost together to demonstrate*], and you look at the top number and it says 100. Well, that doesn't really make sense, so your answer would be the other number." 400

I realized that knowing the 90-degree angle was really important in their predictions about the measurement of other angles. Familiarity with the benchmark right angle allowed the students to make reasonable predictions and then confirm them with a protractor.

CHAPTER

4

Creating and applying definitions

CASE 17
Too skinny, too pointy, going the wrong way
Dolores
Grade 3, January

CASE 18
Three sides, three corners
Andrea
Grade 2, January

CASE 19
Is a square a rectangle?
Natalie (Observer)
Grades 1 and 2, April

CASE 20
Parallelograms or not parallelograms?
Dan
Grade 3, November

CASE 21
What is a rectangle, anyway?
Olivia
Grade 3, October

Students may know the definition for triangle: a closed shape with three straight sides. However, they are sometimes unable to reconcile that definition with the mental image they hold for the term *triangle.* Two comments from second and third graders, each considering a set of triangle pictures, highlight this difficulty.

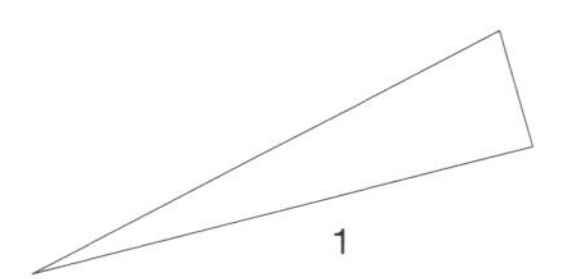

Third-grader Francie wrote, "Shape 1 is not a shape because I know my shapes." About the two triangles (R and L below), second-grader Susannah observed, "I think that they are part of the triangle family, but they're not real triangles."

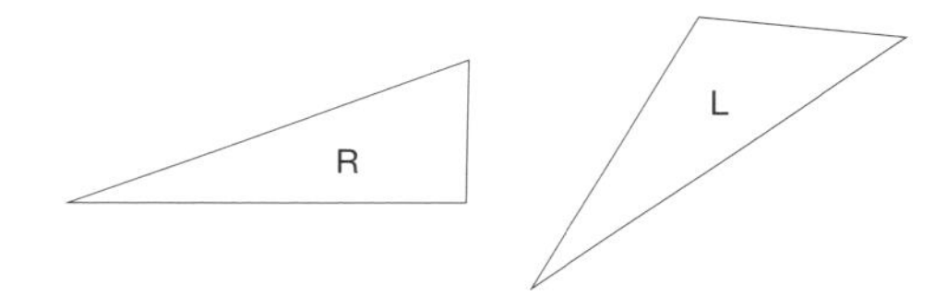

This is typical of the struggle many children have as they try to make sense of the definitions for geometric shapes.

The cases in chapter 4 allow us to examine the way children develop and apply geometric definitions. We see second- and third-grade students studying triangles, and first and second graders who are focused on squares and rectangles. Their comments help us investigate the very purpose of definition. In the last two cases in the chapter, cases 20 and 21, we observe third-grade children developing definitions for parallelograms and rectangles.

As you read the cases, examine these ideas on two levels. First, consider your own thinking:

- What are your definitions for these geometric terms?
- What is the difference between a definition and a list of properties or attributes?
- What is the purpose of a definition?

Then, examine the following set of issues through the eyes of the students:

- What specific issues do they need to consider in order to make sense of definitions for *triangle*, *square*, *rectangle*, and *parallelogram*?
- What is the process they go through as they learn to apply their definitions?
- Looking beyond the specific geometric content of this set of definitions, how do children develop a sense of the purpose of definition?

case 17

Too skinny, too pointy, going the wrong way

Dolores
GRADE 3, JANUARY

For the past several years, I have begun a study of geometry by asking my third-grade classes to identify shapes or to tell me what they know about a collection of simple shapes. Each year there is much confusion. I am always struck by the limited view of triangles so many children have. The majority of students I've worked with feel the only "real" triangles are equilateral triangles, arranged with one side horizontal. Some students can stretch their thinking enough to include a few other kinds of triangles, as long as they are not too far removed from this image.

I have vivid memories of one particular class who held a typically limited view of what makes a triangle. That year we went to dictionaries and to the glossary of our mathematics textbook, where the class found what they called "the rules of being a triangle." That day, the entire class felt both amazed and proud of themselves for expanding their understanding of triangles to include all three-sided closed figures.

Several weeks later I decided to check on their understanding of triangles, only to find that they were right back where they had been before they encountered the definitions. They still considered only the equilateral triangle, oriented in the most traditional way with its point at the top, to be a "real" triangle. I was stunned! What had happened to their expanded view? These results showed they had not moved from their original narrow thinking.

At that point I did something I had never done before. I told them that I had been spending the last few years trying to figure out how children learn about mathematics. I added that I was always trying to get to the bottom of ideas that are hard for kids and why those ideas are difficult. I asked them to remember what it was like to think about the definitions and how that changed or didn't change their ideas. I was surprised when the conversation moved to a higher level. The children were eager to think about my question with me. They were very introspective and articulate.

They spoke about what it was like to try to expand their knowledge of triangles. One student said it was really hard to think of very differently shaped triangles as still being triangles because she felt as though she had really only known "one kind the best," and she had known that triangle her whole life. Another student responded that she felt she really had to break out of the thinking she did before and let some new ideas into her head. I was dazzled! They were really thinking hard about their own learning process.

As their conversation went on, some mentioned the need to take a fresh look at things they thought they knew. Someone suggested, "Old stuff needed to move over in order to fit in new stuff." Some of those students pointed out that the "regular" triangle might always feel "more like a triangle" than some of the others. I remember thinking they were probably right.

With those memories in mind, I decided to start our study of geometry in a different way this year. I didn't provide any shapes for the children to examine. Instead, I simply asked the students to list the shapes they knew and the properties of those shapes. We had just been studying the properties of rocks and minerals, so this term was familiar to them. When I asked them to paraphrase the assignment, they said it was to write the rules of each different shape.

Some wrote things like "triangles have three sides," or "three lines and three corners." A few tried to sketch many-sided shapes such as octagons or hexagons. Others tried to make heart-shaped, diamond-shaped, or kite-shaped figures and name them. I saw that they didn't see the difference between a mathematical term like triangle or square and the familiar name for an object like diamond or heart. In general, they didn't write much about the properties (or "rules") of a particular shape. This assignment helped me see what ideas they were bringing to our geometry work.

Next I drew these shapes:

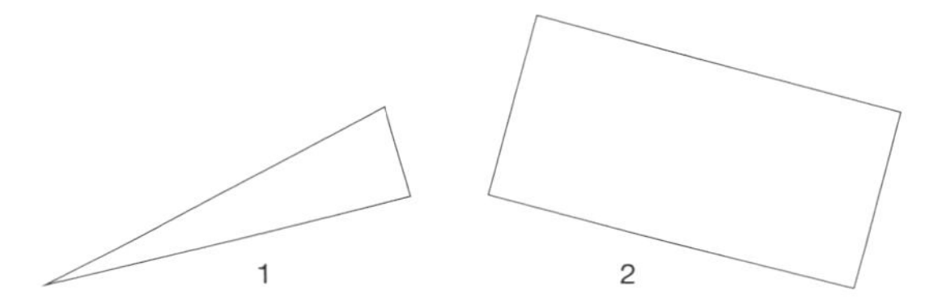

Fig. 4.1. Third-grade students write about shape 1 and shape 2

I asked the class to write in their journals, telling what they knew about these shapes. Most of the children didn't think shape 1 was a triangle. The group was divided about shape 2 being a rectangle. Following are some examples of the student work and what I noted in their writing.

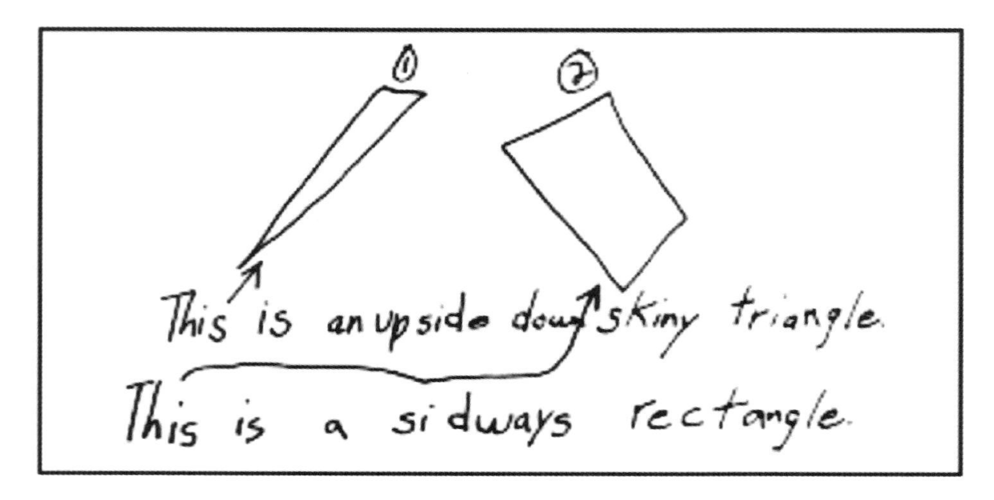

Fig. 4.2. Mei-ying's journal entry

Mei-ying seemed comfortable with the orientation of these shapes. She made note of the upside down and sideways aspects in her comments. While she also noted the feature "skiny," this aspect of the shape didn't prevent her from calling it a triangle. Many third graders I've worked with over the years immediately disqualify this as a triangle because it is too long, too skinny, and too pointy.

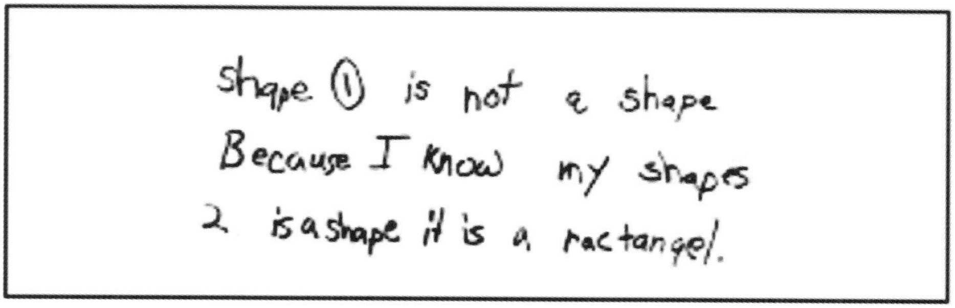

shape ① is not a shape
Because I Know my shapes
2 is a shape it is a ractangel.

Fig. 4.3. Francie's journal entry

Other journal samples showed a limited view of the triangle. Francie wouldn't call the first shape a triangle. I would guess she has a memorized image of a triangle, and this triangle does not match that image. I am reminded of the triangles in baby and toddler toys, such as shape sorters; they are usually equilateral triangles. Perhaps the limited nature of children's early experience with triangular shapes affects how they think later.

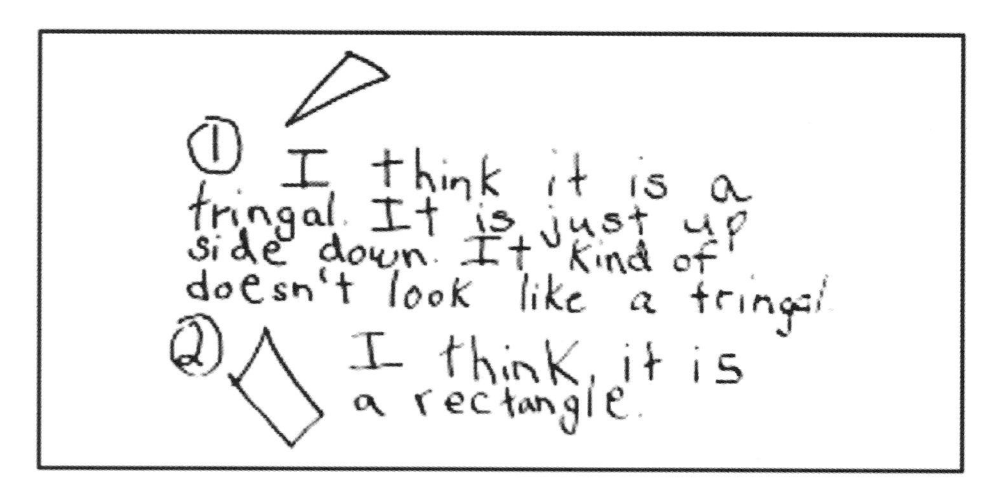

Fig. 4.4 Louis's journal entry

Louis's journal was interesting to me because he made note of the "upside down" aspect of the triangle and wasn't concerned about the turn of the rectangle. I was also struck by what I would call the gut reaction of "It kind of doesn't look like a triangle." Interestingly, Louis drew the triangle much less elongated than I had.

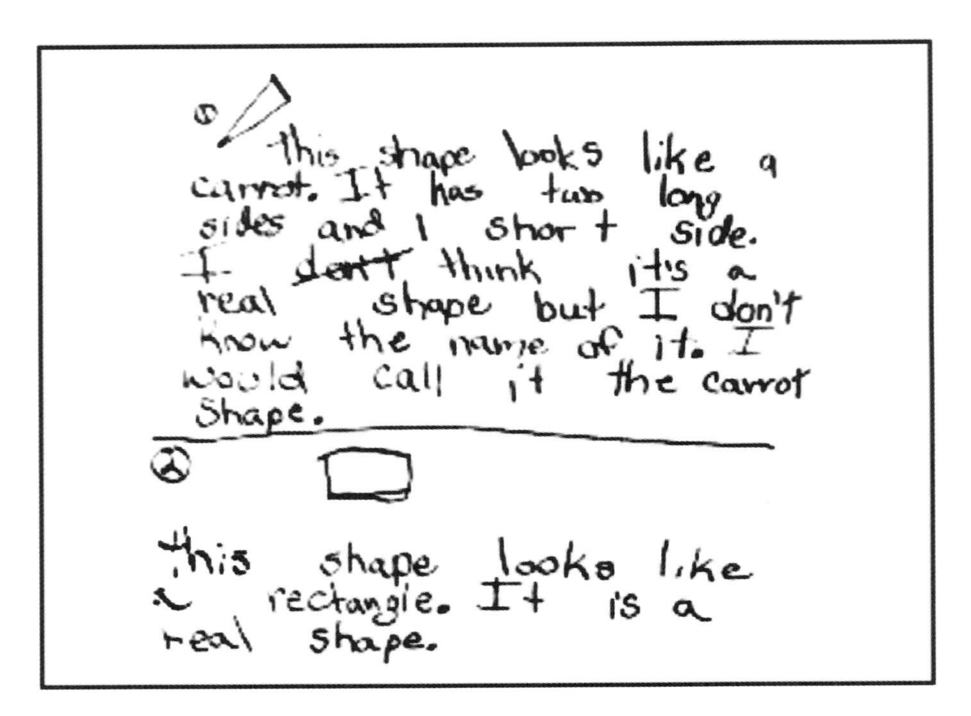

Fig. 4.5. Shannon's journal entry

I saw a different kind of indecision in Shannon's entry. While the third sentence originally began "I don't think it's a real shape," the "don't" is later crossed out. There's that word "real" again, and the word *triangle* is never mentioned. "Carrot shape" was certainly original. Finally, I wondered if rotating the rectangle to a horizontal orientation (as Shannon did when she drew it in her journal) made it feel more real.

While Ricky (see fig. 4.6) had some trouble with spelling, his writing showed no problems with the orientation of either shape or the nontraditional look of the triangle.

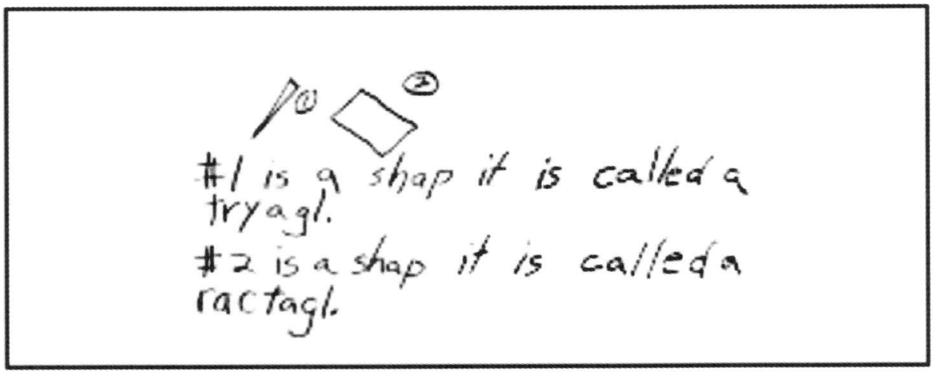

① ②
#1 is a shap it is called a
tryagl.
#2 is a shap it is called a
ractagl.

Fig. 4.6. Ricky's journal entry

As I looked through the collection of responses, I thought about the way I presented this question to them. I had drawn the shapes on paper. How would they have answered if the shapes had been cut out instead of drawn? How would they have responded if they had plastic or wooden shapes that they could move around on their desks? 85

The next day we had a class discussion about whether the shapes were actually a triangle and a rectangle. Again, as in the past, even though the shapes fit the rules (three straight edges and three corners mean a triangle), I could see I was battling an established inner sense. They had a basic belief, and a sense of familiarity, about what is and isn't the best or most "real" triangle. Some students made the case that shape 1 was a triangle by restating the rules. I offered 90 my support by pointing out that "which way it points" isn't a rule. As the class wrestled with whether or not shape 1 was a triangle, I decided one discussion would not change what they believed deep down. Gradually, with exposure to many kinds of triangles, they will become familiar with the whole variety of shapes that we know to be triangles.

case 18

Three sides, three corners

Andrea
GRADE 2, JANUARY

My class has been engaged in a heated debate over triangles. Last year when I had these same students as first graders, they seemed to accept the standard definition of a triangle as a shape with three corners and three sides. This year these same students have been questioning, discarding, refining, and debating what actually makes a triangle a triangle. At times, their perception of what a triangle is can be at odds with the actual definition. To an observer, this may seem like a step backward, but in reality these students are creating a much more solid understanding of triangles. I have been very interested in which characteristics and attributes they notice and how these have led them to refine their definition of triangle.

An intense debate arose during a game of "Guess My Rule," when students were asked to describe the attributes of this group of shapes:

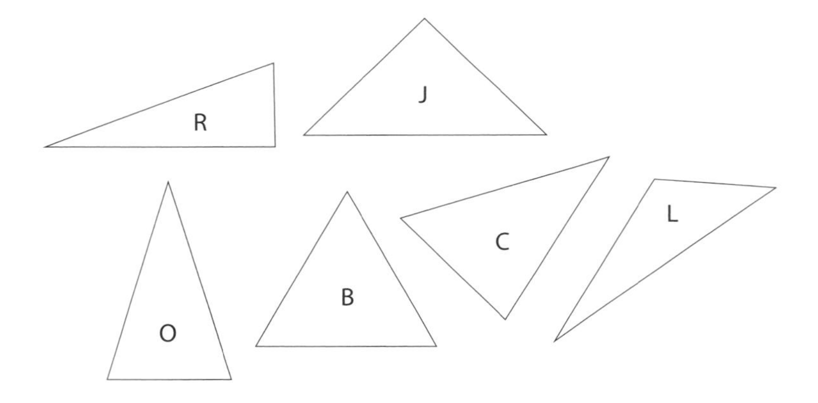

Fig. 4.7. "Guess My Rule" shapes

JACK: I think I know your rule. I think that all those shapes are triangles.

A chorus of voices arose, some in agreement, others adamantly disagreeing. I was surprised by the number of students who insisted that shapes L and R should not be included in the category of "triangles." They had seen many examples of such shapes before and had always been willing to call them triangles.

SUSANNAH: I think that they [shapes L and R] are part of the triangle family, but they're not real triangles.

TEACHER: What makes a triangle a triangle?

NATASHA: Well, I think *tri* means three, so triangle means three.

THOMAS: Three angles!

LAWRENCE: What is an angle, anyhow?

NATASHA: I think an angle is where two lines meet.

TEACHER: Can wiggly lines make angles? [*I hear a chorus of yeses and nos.*] 120

WILL:Yes, I think they can. [*He draws an example.*]

SARA: But no triangle looks like that. You can't have a bumpy triangle. [*She draws what she means.*] 125

NORA: So, I think we need to have a rule that triangles have three straight lines.

KRIS: And I think that their sides have to be equal.

MIKAYLA: No! Look at this one [J]. This is a triangle, right? And these sides aren't all equal. [*Kris shrugs and appears to concede.*]

EVAN: And triangles have to have points. They're pointy, not like curves on the ends. 130

KRIS: Three points.

THOMAS: So all of these shapes are triangles!

SUSANNAH: No! I disagree! The triangles I know look like B, not like L. It's too stretched out. I still think they're just in the triangle family.

The students felt the need to create rules to focus their examination of the shapes. However, 135 some students were not ready to accept these rules, especially if it meant also accepting shapes that felt contrary to their image of what a triangle was. Some of these shapes just weren't "triangle-like" enough. So many of the fundamental concepts and terms in geometry seem heavily based on children's interactions and experiences with their environment.

EVAN: If I'm all stretched out and turned upside down, I'm still Evan. 140

ZACHARY: I think they are all triangles, but I think we think about them in different ways in our head and with our eyes. In our head, we know a triangle has three sides and three corners, but when we see these with our eyes, they don't look right, because we're not used to seeing them like that. But we know with our head that they are triangles. We just have to make our eyes and our heads meet. 145

Wow! I don't know that the other students followed his thinking, but this statement seemed to get at the heart of the debate.

WILL: I still don't think that R is a triangle. It only has one slanty side, and the other two sides are straight. The rest of these triangles have two slanty sides.

The 90° angle made two of the sides look "straight" to him. What is it that makes vertical 150 and horizontal lines appear straight to students, yet diagonal lines do not fall into this category? They are all straight lines; there seems to be something deeper here.

THOMAS: But if you put R together with another one, it would make a triangle. So it must be a half a triangle.

155

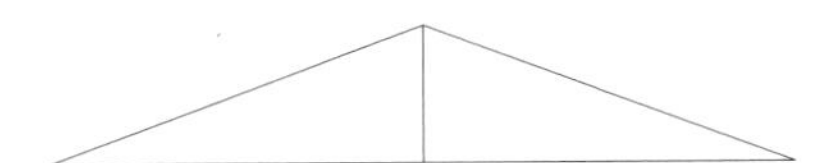

Many appeared to agree with Thomas. This composing of two right triangles to make a larger triangle somehow made each of the original right triangles seem less than "real" triangles.

JACK: I disagree. When we were using the ramps for our motion experiments, they were this shape, and I think they were triangles.

WILL: So look, on this ramp only one side is slanty. So I don't think it's really a 160 triangle. The car can only slide down one side!

LAURIE: But Will, look, it has 1, 2, 3 sides and 1, 2, 3 angles.

WILL: [*pointing to the 90° angle*] Look at this angle here. Look at how this side and this side, they come together like a V. This angle, it makes the sides that come off it go straight. Two straight sides! 165

Again, I think, they're talking about these "straight sides."

ZACHARY: Wait a minute. [*He walks over and turns the block so that the hypotenuse is now lying flush with the desk.*] Look, Will, now there are two slanty sides.

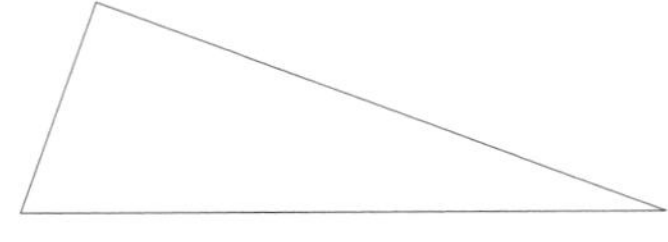

It was subtle, but most seemed to agree that there now appeared to be two "slanty" sides. 170

SUSANNAH: I think we need to add that to the rules. Sometimes you have to turn them around to see, but triangles have to have two slanty sides.

WILL: Two ramps!

LAWRENCE: So you can also turn triangles any which way, and they will always be the same shape. They will always be triangles.

EVAN: Just like you can turn me any way and I'm still me.

The discussion was ebbing and it was time to move on. I asked the students to return to their desks and spend several minutes writing responses to the question, "What is a triangle?" Their written answers revealed a range of responses (see figs. 4.8–4.10), with students focused on a variety of different attributes. This opportunity to explore, debate, and discuss these shapes seemed to have enabled the students to reach a deeper understanding of what it means to be a triangle. Many seemed to have arrived back at the "three sides, three corners" definition, although quite a few were a bit more refined. I did not see as much focus on the straightness of the lines, but many recognized that the sides could be different lengths, and the triangle could be oriented in any direction.

It is clear that our class will continue our discussion and debate on the nature of triangles, but I do wonder: How complete should their understanding of a triangle be at this point? I expect their understanding will continue to grow as they explore other geometric shapes in greater detail.

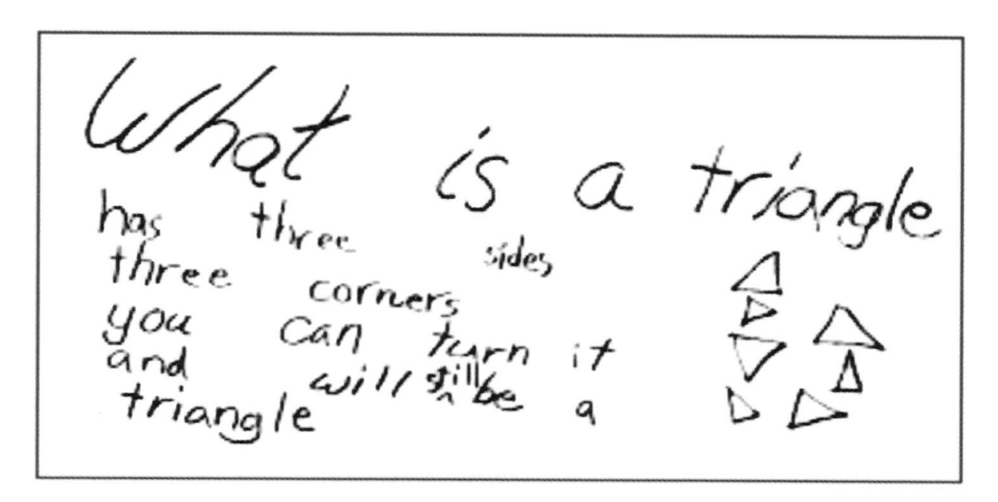

Fig. 4.8. Evan's writing about triangles

It has 3 sides, 3 corners, 2 slants, 1 strate side. If you turn it all around it will still be a triangle.

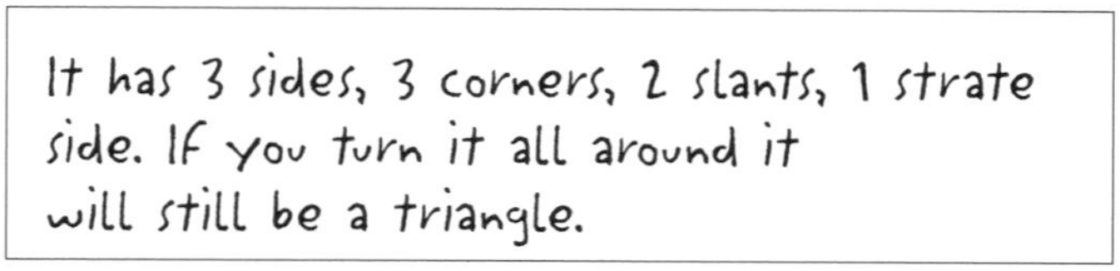

Fig. 4.9. Jack's writing about triangles

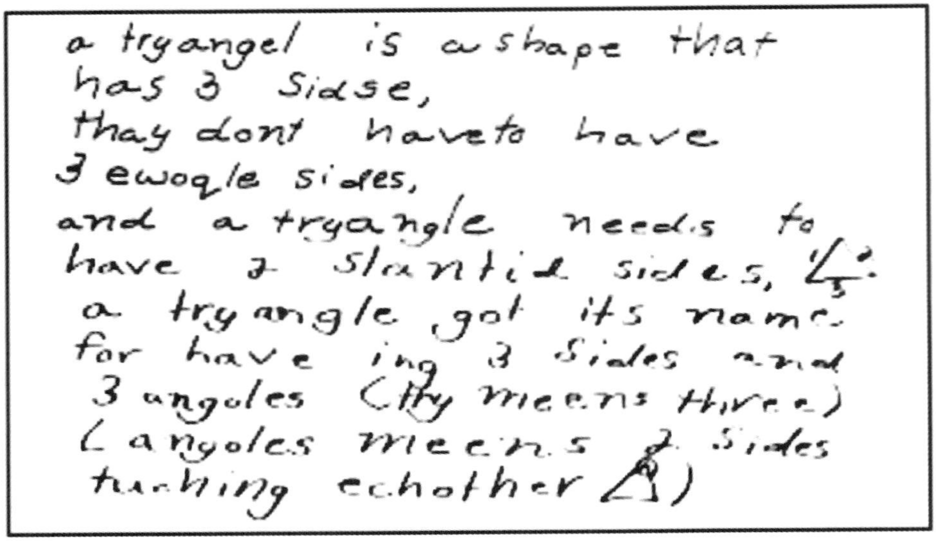

a tryangel is a shape that
has 3 Sidse,
thay dont haveto have
3 ewoqle sides,
and a tryangle needs to
have 2 slantid sides,
a tryangle got its name
for have ing 3 Sides and
3 angoles (try meens three)
(angoles meens 2 sides
tuching echother)

Fig. 4.10. Nora's writing about triangles

case 19

Is a square a rectangle?

Natalie (Observer)

GRADES 1 AND 2, APRIL

I think it's a good idea to know what students do in the year before they come to my third- and fourth-grade class, at least in theory, but in practice I rarely get to visit other classrooms. Due to fortuitous circumstances, I recently had the opportunity to sit in on a math class in a first- and second-grade classroom. The teacher, Ms. Rivera, was welcoming and receptive to my request, even though it was the first day back from spring vacation and not usually a time when you'd want someone observing your class. 195

She sits with a group of enraptured first and second graders, reviewing the work they did just before spring vacation about rectangles. To start, she reads several children's definitions and one rectangle poem. There is an audible buzz from the class. Children are remembering their own definitions, comparing theirs to the ones Ms. Rivera was reading, sometimes asking to have their definition read aloud. 200

GABRIEL: They have four sides and four corners.

MS. RIVERA: In our last conversation, some of you noticed that the paper you worked on was a rectangle. 205

CLARISA: [*trying to add onto Gabriel's comments and jumping in before Ms. Rivera was finished speaking*] . . . and four angles. Angles are where sides meet.

MS. RIVERA: Last time we talked, Luke said angles could move. [*She holds her palms together and fans out her hands to remind them.*] 210

As the children talk, Ms. Rivera writes down the characteristics of rectangles from their conversation: four sides, four corners, four angles.

MARYANNE: Two sides are longer and two sides are shorter.

FINN: Two squares make a rectangle.

A couple of the children demonstrate that the rectangles drawn on the board all have squares in them. All the children (or at least the ones who are contributing to the conversation) seem pretty convinced that rectangles are made up of two or more squares. 215

MS. RIVERA: So, what's a square?

I love this question because I can see Ms. Rivera calmly and serenely putting a question to them that will force them to rethink and reevaluate their working definition. 220

ROBERTO: Four sides, four corners, four angles, and it's a square.

At this point, "it's a square" seems to be part of the definition in the same way that "four sides and four corners" are. *Square* seems to be a characteristic or attribute of its own. Ms. Rivera writes a list of what makes a square a square: four sides, four corners, four angles. 225

CLARISA: Actually, you don't need to say four corners and four angles; they're the same thing.

MS. RIVERA: Look at what we said about squares and what we said about rectangles!

Then Charlie, who had been sitting away from the group most of the time, called out.

CHARLIE: Hey, they're the exact same thing! I'm thinking of shapes with the same definition, but they're not a square or a rectangle. 230

MS. RIVERA: Can you draw what you are thinking of on the board?

Charlie draws a chevron and a trapezoid.

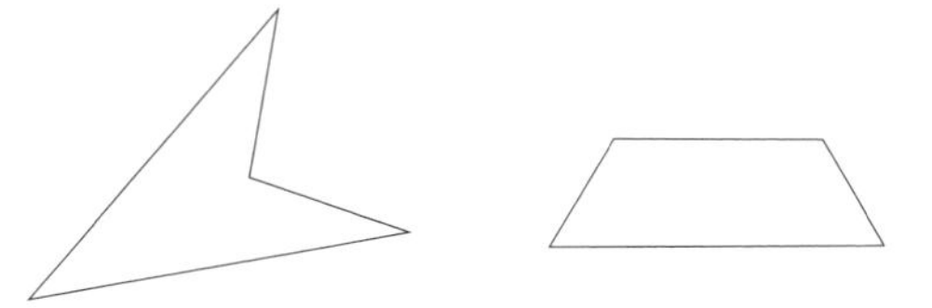

The children seem really intrigued by this new way of looking at shapes alongside the definitions. A bunch of them go to the board to experiment with other shapes. Ms. Rivera focuses their attention on the issue of making a definition. 235

MS. RIVERA: How can we make a definition for rectangles that is more specific?

The children know what she means; because their lists for rectangles and squares are the same, they need to clarify and differentiate between them. 240

MARYANNE: I don't think Roberto's [definition] makes sense. Can I come up and draw it? Someone could think it was like this. [*She draws a square, oriented so it looks like a diamond.*] So we need to explain it better.

MS. RIVERA: Plus, you guys said that a square was not a rectangle.

BRETT: If you turn a square over, it would look the same. 245

Brett attempts to make a distinction between the definitions of a square and a rectangle. He gets an orange pattern block, which is a square, and turns it over from one side to the next, like a square wheel. Every time he turns the square, it looks the same. His point is that this characteristic distinguishes it from a rectangle, which looks different from one 90-degree rotation to the next. He and some other children experiment with different pattern blocks. 250

HELGA: All the sides of a square are the same length.

MS. RIVERA: So what if you had never seen a square before and you looked at these shapes. Would you be able to take our definition and know which shapes were square and which ones were not?

I can no longer resist entering this discussion. Having had similar conversations with my third and fourth graders, I now wonder what first and second graders think about angles. So far, their comments have been mostly focused on the sides or the outside of the shape, not the space within the shape or the shape itself! I go to the board and draw a rhombus—a parallelogram with all four sides the same length—that is not a square. I ask whether this would be considered a square if they used their definition. Some kids call out yes while others say no.

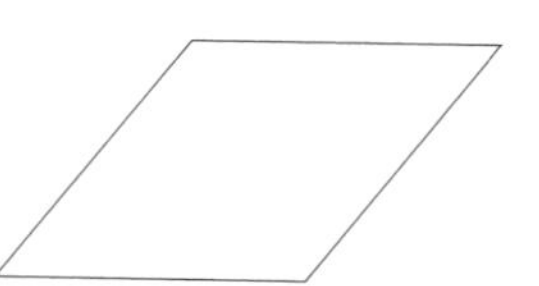

BRETT: If that were a true square, it would go straight up. This one goes diagonally up.

Children continue to offer their ideas, and eventually all agree that squares have "angles that are the same."

The discussion about squares and rectangles ran longer than Ms. Rivera had planned, but she allowed it to continue because so many children were contributing and working through their ideas. Although the distinction between a rectangle and a square was not completely resolved, some progress was made.

I was interested in the way Charlie showed that a variety of shapes would fit the given statements about squares and rectangles. He could focus on the specific attributes of four sides, four corners, and four angles, and yet create a wide variety of shapes. He seemed to be getting at the need for another term, *quadrilateral.* But while Charlie and his classmates noticed that there weren't any squares or rectangles represented on the board, they didn't seem to mind, and no one made any move to narrow their definition of rectangle or square.

It also didn't matter to most children that the description for square was the same as the description for rectangle. In fact, Roberto offered as part of the definition of a square that "it's a square." This group is just beginning to figure out the difference between listing attributes of a shape category and making a useful definition. Brett's action of rolling the square offered an interesting possibility. He was on his way to proving that "square-ness" had to do with looking the same no matter what side you turned it on; he was trying to make a distinction between a square and a rectangle. (The fact that squares are included in the category of rectangles would come later.) If he could figure out what made a square unique, then there would be a purpose to having a definition. Why else would you need a definition?

case 20

Parallelograms or not parallelograms?

Dan

GRADE 3, NOVEMBER

Over a period of several days, my third-grade class had been looking at the characteristics of pattern blocks.

During this discussion, the idea of parallel lines came up, and the word parallelogram was used. Eventually we came up with a working definition of parallel sides: if the sides were extended, the lines would never touch, like train tracks. Damien stated that he thought the red trapezoid was a parallelogram because it had parallel sides and an even number of sides. I wanted to determine if he would hang onto, change, or expand on those two descriptors in his definition of a parallelogram if I gave him a slightly modified, unfamiliar shape. I drew this pentagon on the board.

I chose this shape because it has two sets of parallel sides rather than one, and has five sides rather than four. Would Damien consider this a parallelogram? In fact, he insisted it was not a parallelogram because it had an odd number of sides. No one disagreed with him. There was some difference of opinion about whether the yellow pattern block (hexagon) should be considered a parallelogram. They all agreed that it had parallel sides. If we used the defining characteristics presented by Damien earlier—an even number of sides, and parallel sides—then this shape would be considered a parallelogram.

So, what is a parallelogram? It is defined as a four-sided figure having two pairs of parallel sides. A consequence of this definition is that both sets of opposite sides are parallel and equal in length. Because the definition the class was developing was not in agreement with the actual definition, I was unsure of how they would come to the standard mathematical definition. I decided to give the students a homework assignment so they could work on this idea. At my instruction, they traced the orange square, blue rhombus, and tan rhombus on the top of a journal page and labeled them parallelograms. On the bottom, they traced the yellow hexagon, red trapezoid, and green triangle and labeled them not parallelograms.

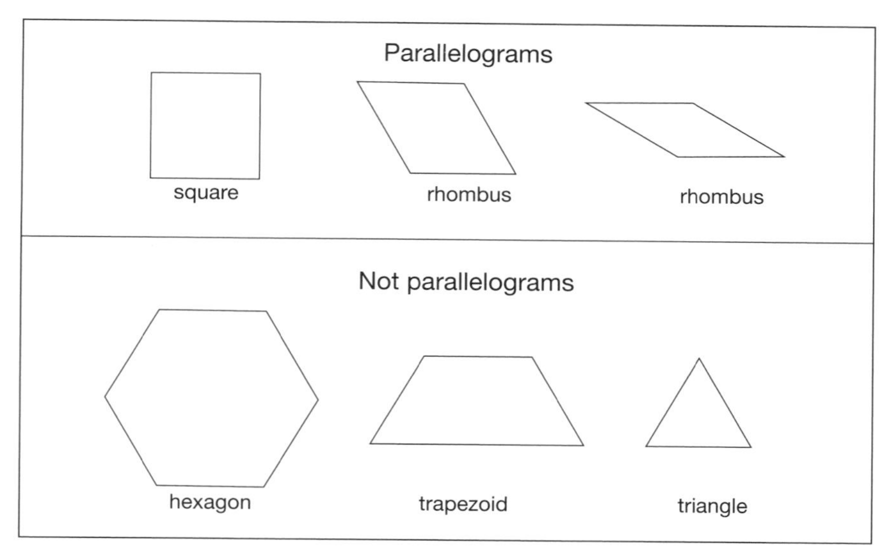

I explained that mathematicians had agreed which shapes were parallelograms and which were not. I told them that three of these shapes were what mathematicians would label parallelograms and the other three were not. I asked them to write about what was the same about the top three and what was different about the bottom three. I hoped that this assignment would lead them to question the completeness of their current definitions. Following are some of the children's responses to the assignment and my reflections on their thinking.

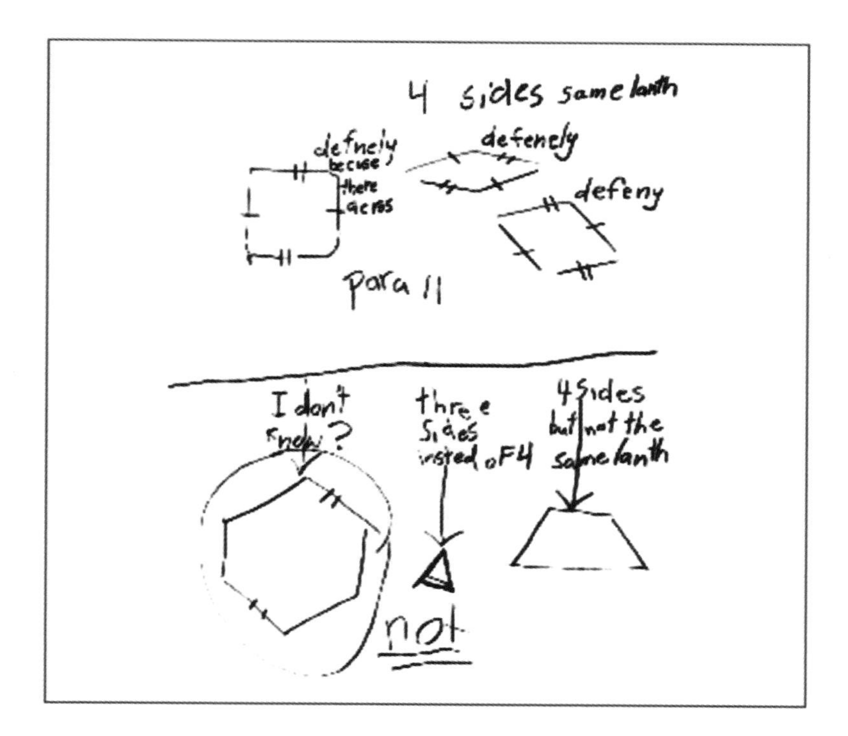

Fig. 4.11. Eleanor's ideas about parallelograms and "not parallelograms"

On the shapes that are in the parallelograms section, Eleanor (see fig. 4.11) has marked opposite sides in a way that suggests they are equal in length. (These symbols had been used in class.) She also wrote "4 sides same lanth." For these particular shapes, her observations are 320 accurate. However, nowhere has she made a comment about parallel sides (I think the word "parall" she wrote is the label for the section of shapes rather than an observation she has made). Unfortunately, there were no parallelograms with adjacent sides of differing length to challenge her comment about four sides of equal length, so I don't know how she would have explained those shapes in this section. That is something I'd like to talk with her about. 325

For the "not parallelograms" category, Eleanor clearly noted that the triangle has three instead of four sides and the trapezoid has two sides that are not the same length, even though there are four sides. These comments seem to suggest she recognizes that parallelograms must be quadrilaterals and thinks they have "4 sides same lanth." For the hexagon she wrote, "I don't know?" She does not comment about the six sides. 330

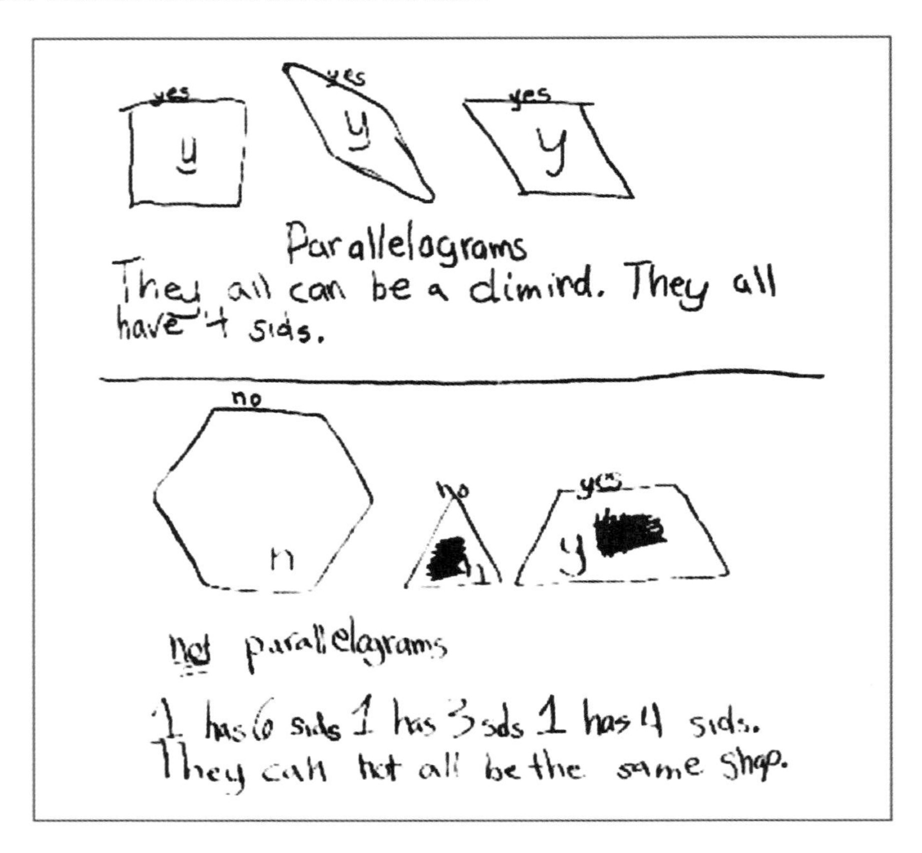

Fig. 4.12. Ally's ideas about parallelograms and "not parallelograms"

Ally (see fig. 4.12) noted that the top three shapes all can be diamonds and they all have four sides. However, she makes no comment about the parallel sides or side-length characteristics of these shapes. When comparing the bottom shapes with each other, Ally pointed out that one has 335 six sides, another three, and the third four. She used this point to say they cannot be the same shape. While this observation distinguishes them from each other, it does not rule out the trapezoid as a parallelogram. Ally does not give any explanation for why it is not in the top group. She has made accurate observations, but has not really articulated a clear understanding of why these shapes are categorized as they are. 340

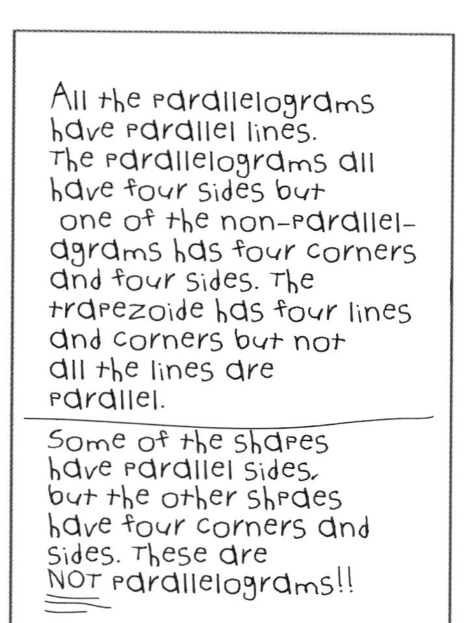

Fig. 4.13. Christian's ideas about parallelograms and "not parallelograms"

Christian (see fig. 4.13) seems to have the deepest understanding of what defines a parallelogram, although some of his comments are a little confusing. He notes that all the parallelograms have parallel lines and four sides. He also explains that while one of the nonparallelograms, the 345 trapezoid, has four corners and four sides, not all of the lines are parallel. This seems to clearly include two opposite pairs of parallel lines and four sides as necessary characteristics for a shape to be labeled a parallelogram. Christian's comment that "some of the shapes have parallel sides but the other shapes have four corners and sides" seems to be a way to explain why the hexagon does not fit the description of a parallelogram. 350

So what did I learn? The understanding of what constitutes a parallelogram is much more complicated than I had thought. It feels as if the majority of the class was moving toward an understanding of "four-sidedness" and "opposite parallel sides" as defining a parallelogram. However, some children were clearly having difficulty keeping both of those characteristics in mind at the same time. I also see that using only the pattern block pieces as examples can lead 355 them into thinking that all four sides of a parallelogram have to be the same length. I wonder if that is a coincidental observation by Eleanor, or if she includes that as part of her construct of what defines a parallelogram? Is length the most distinguishing characteristic for her? If so, perhaps that is why she is confused about whether a regular hexagon fits into the parallelogram category. 360

These comments by the children help me see some of the key variables that I need to explore with the class. We will need to look at a wide range of examples (e.g., adjacent sides of different length; squares and rectangles as special kinds of parallelograms) and counterexamples (shapes that fit the students' definition but are not parallelograms) to help clarify the actual defining characteristics of parallelograms. 365

case 21

What is a rectangle, anyway?

Olivia
GRADE 3, OCTOBER

In my work as a math resource teacher, I was preparing to introduce some geometry concepts to a group of third graders. I knew that this group had had limited exposure to the math ideas in geometry, so I began by asking them to look around and tell me what shapes they could find in the classroom. Many eager hands went up and a lively discussion ensued. 370

While they identified a number of different shapes, including a trapezoid (table), a sphere (the globe), and a cylinder (the supporting post), most of the shapes that students pointed out were rectangles or squares. I noticed early on that anything vaguely "square-like" was called a square, including books, the refrigerator door, and computer screens. Only the refrigerator door was actually a square. They did label as rectangles the maps, the chalkboard erasers, the windows, and the doors on the closet. I wondered how these students were deciding which to call squares and which to call rectangles. 375

TEACHER: I noticed that you called the maps and the walls rectangles and the computer screens and the refrigerator door squares. Why did you decide that some were squares and some were rectangles? 380

LINETTE: Rectangles are longer and skinnier than squares.

KAMALI: Yeah. A square is shorter and fatter.

TEACHER: Do other people agree with what Linette and Kamali said?

The response was many nods, yeses, and other expressions indicating agreement. I drew a long skinny rectangle, a square, and a more "square-like" rectangle on the board and asked the students to identify them. They all agreed that the first shape was a rectangle and the last two shapes were squares. I asked the class what the shapes they called squares and the shapes they called rectangles had in common. 385

DANIEL: They all have four sides and four corners.

Again, much agreement was indicated. 390

TEACHER: Daniel's right. Mathematicians say that squares are a special kind of rectangle. Linette is a person, but she's a special kind of person. She's a girl. Lamar is a person, but he's a special kind of person, a boy. What do you think makes a square a special kind of rectangle?

395

The students answered again that rectangles were longer and skinnier, whereas squares were shorter and fatter. In retrospect, I'm wondering if the mostly horizontal orientation of the rectangles they had identified was affecting their descriptions. What would have happened if the long, skinny rectangle had been turned on its side? At the time, I kept the discussion focused on squares. 400

TEACHER: You all agreed that squares and rectangles have four sides and four corners. Is there anything special about the sides of a square?

Among the murmured responses, I heard Elizabeth say that the sides of a square are "even."

TEACHER: Elizabeth said that the sides of a square are even. What do you think she means by that? 405

MATTHEW: All the sides are the same.

Lots of students agreed with Matthew and Elizabeth, but I remained far from convinced that this question had been settled. What is the relationship between these two types of geometric shapes? How do these shapes fit in with other more general shapes?

I decided to focus our next lesson on quadrilaterals. I had a set of shape cards, each with a 410 drawing of a 2-D shape. For this work, I used only the cards with four-sided figures. To set up a game of "Guess My Rule," I told the students to think of a rule that some of the shape cards followed and some did not. After a few minutes, Quetcy [pronounced Ket-see] raised her hand, and I had her whisper her rule to me: "Shapes that look like a box." According to the rules of the game, Quetcy began by choosing one shape that met her rule and one that did not; she 415 showed these to the class. The rest of the students took turns picking a shape card and placing it in one of two piles: "fits the rule" or "does not fit the rule." After each placement, Quetcy either agreed or moved the shape card to the correct pile. When all the shapes were placed in one pile or the other, we tried to guess Quetcy's rule.

A number of students offered "shapes that have four sides," but then they noticed that all the 420 shapes fit that rule. Suddenly Ayesha became very excited, waving her arm to be called on.

AYESHA: All the shapes that follow the rule are squares and rectangles.

KAMALI: The other shapes would be really funny-looking boxes.

TEACHER: What is it that makes them look funny?

GISELA: They're slanty. They're not straight like the others. 425

LINETTE: I know that all the shapes that follow the rule aren't squares, but something seems "squarish" about them.

I was hoping that Linette would say something more about her idea, but she didn't, and no one else picked up on it. While the students seemed to think that these shapes were somehow "straighter" than the others, I wasn't sure if they were noticing the corners or the angles. With 430 more questioning, would they see a difference between the angles in these "box-like" shapes and the angles in the nonrectangular quadrilaterals?

I chose the following four quadrilaterals from the set of shape cards: shape A, a rectangle; shape E, a trapezoid; shape F, a square; and shape P, a quadrilateral with no parallel sides.

435

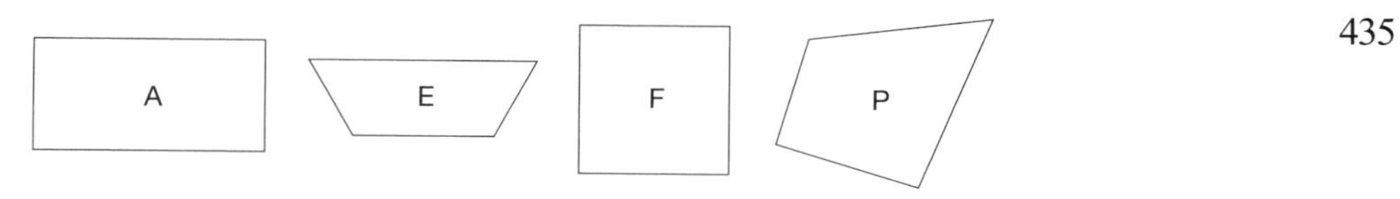

I included a square because I was curious to see if the students would identify it as a rectangle; since the beginning of the unit, we had called a square "a special kind of rectangle with all sides equal." I gave each student a sheet with the four shapes pictured at the top. The directions read as follows:

440

> Look at the shapes above and answer the following questions:
> 1. Which shapes are rectangles? How can you tell?
> 2. Which shapes are not rectangles? How can you tell?
> 3. What do all four shapes have in common? In other words, what is the same about them?

I hoped that asking the students to focus on and write about these four quadrilaterals might serve two purposes: (1) to show me where their thinking was on the question of why some were rectangles and some were not, and (2) to push them, at the same time, to think really hard about this question. I wanted this writing to advance their understanding.

As usual, I was struck by the seriousness with which these students set to work answering my questions. I could see by the expressions on their faces how difficult it was for them to express, in writing, what distinguished these shapes from one another. Despite the difficulty, when math class was over, all but two of the students had finished the assignment. 445

The end of the day came quickly. With all the papers in hand, I spent the five minutes I had before bus duty looking over what the students had written. My initial reaction was disappointment. At first glance, it didn't appear that the students had progressed in their thinking. For instance, Quetcy still used the criteria of "looking like boxes" to categorize which of the shapes were rectangles, and she didn't spell out what she meant. 450

Later I had more time to review their work. What did these papers show that the students knew? One thing struck me immediately: All but two of the students had correctly identified shape F, the square, as being a rectangle. Veronica, Daniel, and Lamar all wrote about corners as a factor in determining which of the shapes were rectangles. Veronica wrote that shapes E and P are not rectangles "because they don't have equal corners." Daniel wrote, "E and P are not rectangles because the corners are different." 455

I also think that the students who wrote about "straight" and "crooked" lines were on to (460) something important. They were clearly not assigning the same meaning to these terms as I do. However, in looking at their writing and reflecting more on our earlier discussion, it is apparent that their idea of straightness and crookedness may refer to the relationship of the lines to one another. Kamali's drawings make me think she is using the word "criket" (crooked) to mean "not perpendicular." (465)

What are the areas of confusion shown by their work? Ayesha and Gisela both wrote about rectangles needing to have sides that are equal or the same. Do they mean two pairs of equal sides? Neither of them recognized that shape E does in fact have one pair of equal sides. Ayesha, Linette, and Nathan wrote about rectangles being defined by having two long and two short sides, yet all three of them identified shape F, the square, as a rectangle. Were they just (470) repeating back what they had been told about squares being a special kind of rectangle with no real understanding of what that means? Was that also true of other students? I now think they need to recognize the special angles that squares and rectangles have in common in order to really understand why a square is a kind of rectangle.

There were other interesting ideas. Katelyn seemed to be playing with the idea that being (475) able to fold a shape along a line of symmetry to form congruent halves is one proof that the shape is a rectangle. She said about shape A, "I can tell because if you fold it, it will match." However, when she wrote, "Shape E is not a rectangle because the sides are not matching," she failed to see that shape E also has a line of symmetry. Juan continued to show an interest in the space inside when he wrote of shape A as being a rectangle because it has "anof space." (480)

What implications does all of this have for instruction? As I ponder our earlier discussion and the students' writing, I think I held back information that would have been useful to them. As some of the students began to focus on "corners that are the same" and "straight" versus "crooked" lines, weren't they in fact beginning to think about right angles and parallel and perpendicular lines? I wish now that I had put those definitions out there. I realize that not every- (485) one in the class would have been ready to grasp those concepts. Others, however, might have come to a heightened understanding of what makes some quadrilaterals rectangles while others are not, and why squares have certain characteristics that place them in the class of figures we call rectangles. I don't mean to say that I would have just told them that rectangles (and squares) have two sets of parallel sides and four 90° angles. However, by introducing the concepts of par- (490) allelism and right angles, I would have given the students another frame of reference for their examination of four-sided figures.

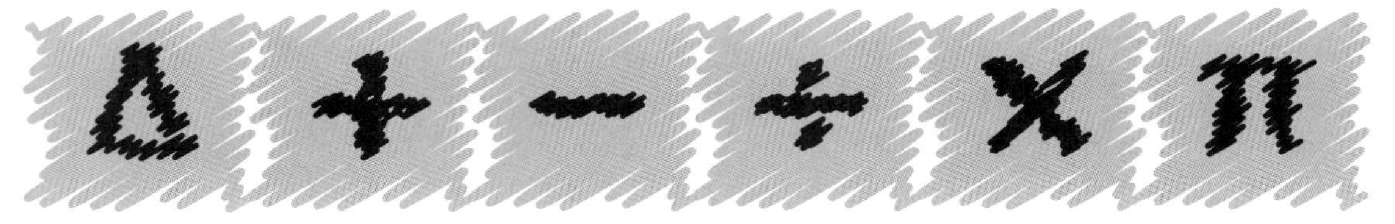

CHAPTER

5

Comparing shapes

Suppose someone showed you shapes A, B, and C, and asked, "Which of these shapes are the same?" This seems like an easy question. Take a minute to think about how you would answer it.

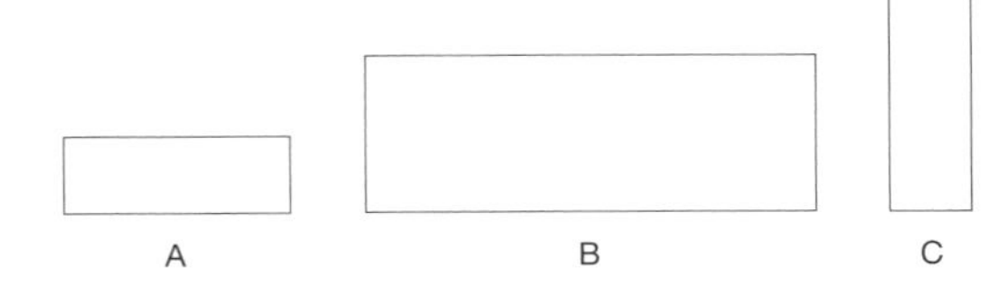

Would you say they are all the same because they are all rectangles? Would you say figures A and B are the same because one is just an enlargement of the other? Would you say that figures A and C are the same because they are the same rectangle pictured in different positions? Exactly what do we mean by the same, by similar, by congruent?

In this chapter, the children are working on these questions. Josie's fourth graders construct polygons that are similar to each other. Dolores's third graders and Ellie's first graders explore the idea of congruence as they build shapes out of squares and triangles and then try to determine which shapes are unique and which are not. In Sally's case, we examine the thinking of second graders as they confront the issues of orientation: Are rectangles like A and C in our example the same, or are they different?

case 22

Exploring similar shapes

Josie
GRADE 4, APRIL

My class had done a unit in geometry earlier in the year in which they had worked with a variety of 2-D figures. From that work, my students knew the names and shapes of the familiar polygons. Now we were beginning another geometry unit, one in which we would look at the relationships between shapes. 5

I started this activity by displaying three similar triangles on the overhead projector. I asked the students what was the same and what was different about these shapes.

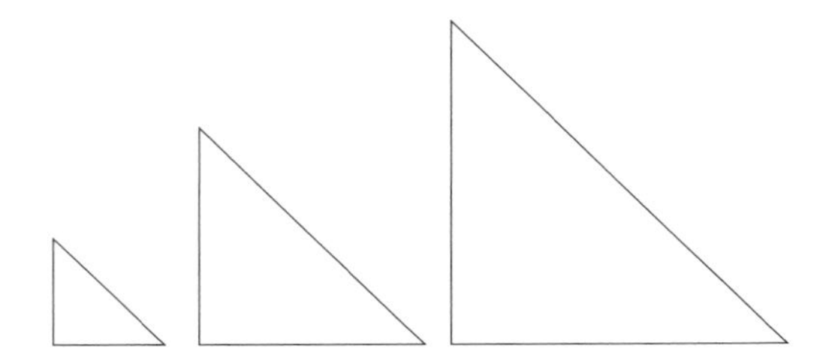

I was pleased with the responses they offered. They listed these things as the same: The shapes all have three sides, one right angle, and two acute angles; and they are all triangles. Their size 10 was mentioned as being different. Some talked about the length of the sides as being longer or shorter; others just referred to the size of the whole triangle. I was even more pleased to hear Deb's comments.

DEB: They are all the same.

TEACHER: How can that be? 15

DEB: Each one is like the smallest one, but they have kind of been stretched.

After discussing Deb's "stretched" concept, I introduced the word *similar.*

TEACHER: The triangles are similar because they are "stretched" in a way that keeps the original shape's proportions. [*I held up a rectangle from a set of plastic shapes.*] Can someone show us how to make a larger rectangle that would be 20 similar to this one?

After some discussion, the class agreed that one similar rectangle could be made with four of the original rectangles. The height of the new one would be twice the height of the original, and the length of the new one would be twice the length of the original. From there, the students quickly decided that to make the shape similar, the sides all had to be changed in the same way 25 (for instance, all sides doubled in length). After these comments, they had no trouble determining that with nine rectangles, they could make another similar shape.

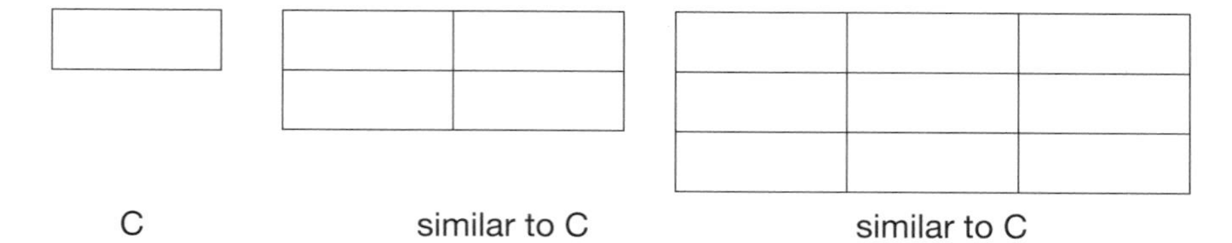

With this preparation, I directed the class to the day's worksheet. The assigned activity would be similar to what we had just done, but would extend the work to shapes other than 30 rectangles. They were to take a polygon shape and build a larger figure similar to it. I distributed buckets of plastic shapes for them to use. The set included a variety of polygonal shapes, including the six pattern block shapes. In order to build the similar pairs, students could use any combination of these shapes.

While I watched them work on this assignment, I noted that some students worked quickly; 35 others made mistakes and had to redo some of their examples. Everyone had difficulty with the hexagon. As I watched, I tried to figure out what aspects of the shapes my students were paying attention to as they did this work. What were they recognizing about similar shapes? How did they use that information to work on the problems?

I spoke to Robby, a student who usually does very well with visual tasks, but who was now 40 struggling with the hexagon.

TEACHER: What is the problem with this?

ROBBY: There is no way to make the next shape with the hexagons. It won't be doubled . . . it would have to be larger . . . like five or ten times larger. I always get a honeycomb shape. 45

TEACHER: This is a challenging shape. Maybe you need to use a different strategy for this one. I will tell you that it is possible to do.

I watched Robby become more and more frustrated as he struggled with the hexagons and finally decided to give him a hint.

TEACHER: Do you think it would be possible to make a similar hexagon if you used a 50 shape or shapes other than the hexagon itself?

A look of relief came over his face. I realized then that he didn't think this was allowed because we used only rectangles when we did the rectangles. He selected the trapezoid and went to work. Later when I asked him how he decided to start with the trapezoid shape, he said it was because I told him to. When I reminded him that I did not select the shape, he stated that 55 he must have chosen it because he knew it was half of a hexagon. He was successful in just a few minutes. As Robby worked, it appeared to me that he was able to think of both the original shape (the hexagon) and the concept of doubling the length of its sides.

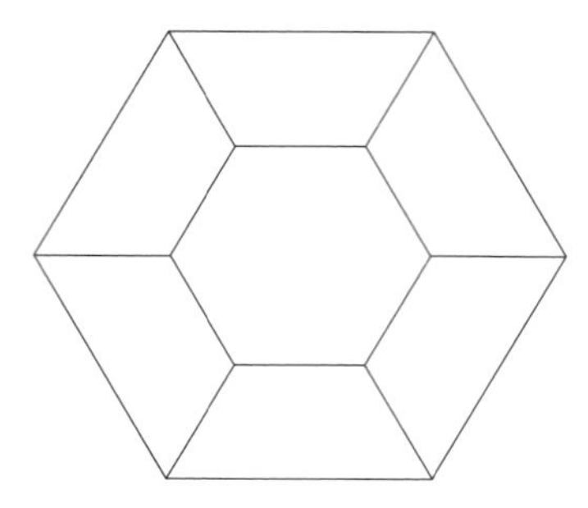

Rose was also struggling with the hexagon and needed me to suggest trying a different strategy. I asked her, “If it cannot be done with just hexagons, then what can it be done with?” This was all she needed. Rose immediately began to consider using another shape. She selected the rhombus, took a handful of them, and tried a couple of different placements of this shape around one hexagon. Soon she too was successful. 60

65

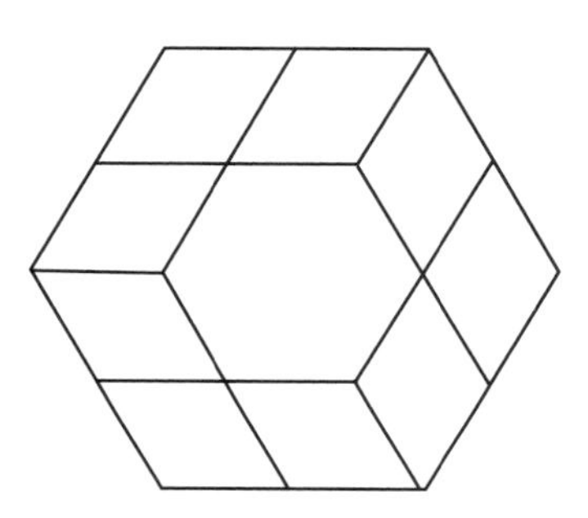

Paylee generally needs my assistance and a bit more time than others when facing a math concept that is new to her. She wasn’t sure how to work with the square. As soon as Paylee admitted she didn’t know what to do, I reviewed the rectangle procedure with her and suggested that she work with the square in the same way. She was successful with that shape, telling me that the sides of the next largest square had to be two sides long, because the original square was only one side long. Paylee used the first square as a measurement tool to test her larger square, lining it up against one edge of the larger figure and sliding it across, counting “1, 2” to show that the top of this similar square was equal to two lengths of the original square. She checked the height in the same manner. It was interesting to see how concrete her measuring was. She literally wanted to see that the sides of the new square were exactly the same as two of the sides of the original square. 70 75

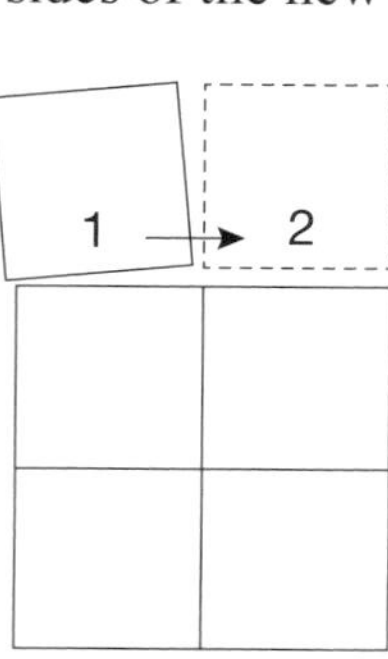

I stayed with Paylee while she tried the hexagon. She started by putting hexagons together and quickly stated that this was not like the others. I asked her what she meant, and she said that it couldn't be put together like the others because it didn't fit like the others. It had spaces. I suggested that she try other shapes. When she didn't appear to know where to start, I separated a group of trapezoids and rhombi from the rest of the figures on her desk. She took the hint and picked up a rhombus. When this wasn't successful, she then tried using the trapezoid. Here are some of the arrangements she tried:

I asked Paylee what she was trying to do, and she stated that she was trying to make each side twice as long. She worked without paying attention to the shape as a whole and therefore didn't know a particular plan wasn't working until she was well into it. I tried to help her by focusing her attention on a single line segment and asking her how she could make this side twice as long. She was able to do this with a rhombus.

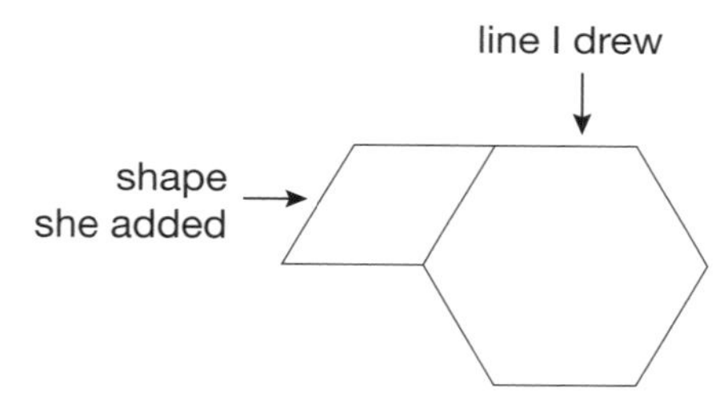

I then asked her what side she would lengthen next. She identified a side, and I drew a line to indicate it. She added a hexagon to her figure, and now had two sides of the larger hexagon correct.

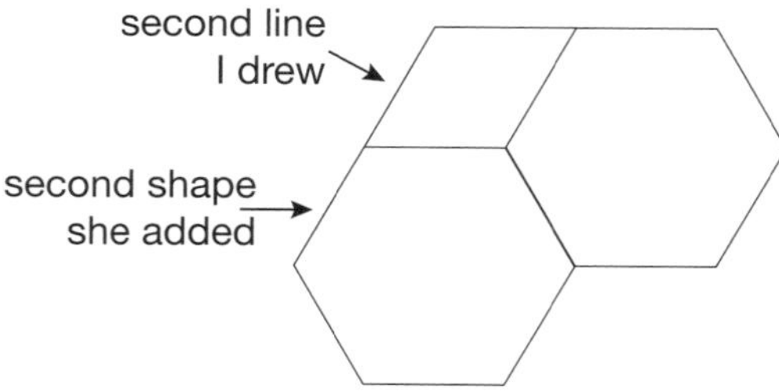

Paylee then added a third hexagon.

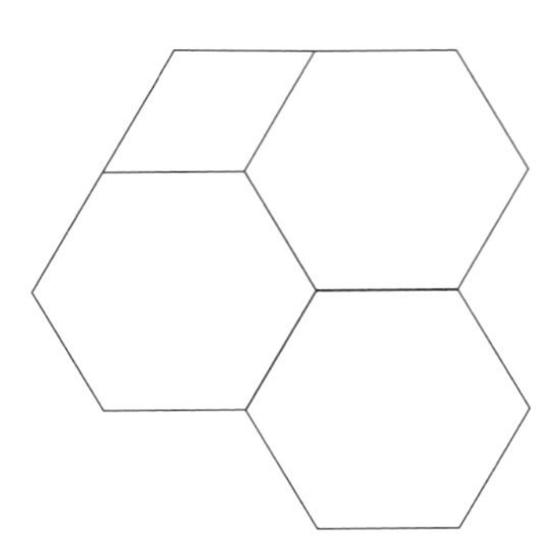

I could see how this would lead to a solution; I could visualize the two missing rhombi. But Paylee couldn't. When I asked her what shape she wanted to use now, she tried the trapezoid. She could see that it was wrong, and finally tried the rhombus shapes. She recognized that this worked, and completed her larger hexagon with all the sides doubled in length.

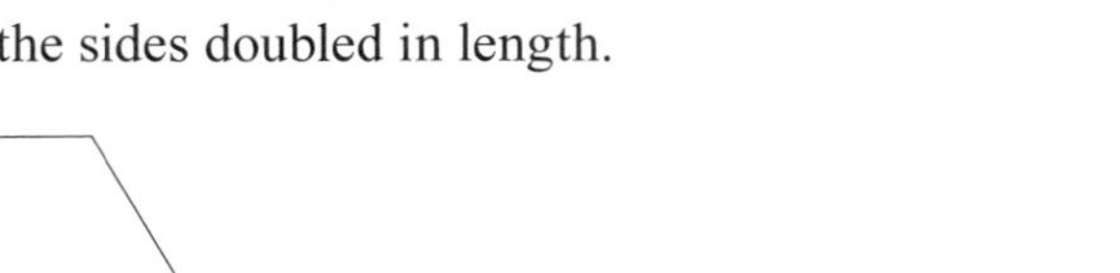

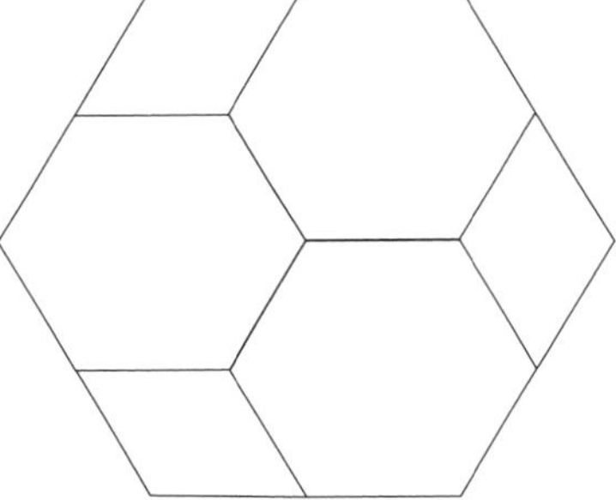

I was left wondering why she didn't see this when I expected her to. I thought it was so obvious. How do I help her recognize the space? Did concentrating on the length of the sides take away her perception of the whole shape? I would need to think about this for a while.

It interested me that with a hexagon, they all solved the problem in a different way. Rob and Rose found the task much easier than Paylee. Paylee not only needed a lot of guidance from me, but at the end of the task, it didn't appear that she really understood. She apparently understood that she needed to double the length of the sides, but didn't have the strategy to do this. I also noticed that all three were focused on the side lengths rather than noticing the angles or the whole shape of the hexagon. They started with the parts of the hexagon and worked toward building the whole shape. What does this tell me about their thinking? I did ask all three students what they thought about as they did this work. They all said that they thought about the shape itself and how to make the sides longer. I am left wondering what other ideas about similar figures we could work on this year.

case 23

Making shapes with triangles

Ellie

GRADE 1, NOVEMBER

During our previous lesson on shapes, the discussion had a strong focus on the triangles found within different shapes. I thought I would pursue the idea further by having students build shapes from triangles. There was one rule: The sides of the triangles must touch completely. We discussed what it meant for sides to touch completely, and as we talked, the students moved the triangles around, experimenting with various placements to decide which arrangements followed the rule. After I felt confident the children clearly understood the rule, I sent them off with their partners to make as many shapes as they could with the equilateral triangles. 120

I watched as each group got started. They all were very focused and engaged. I checked in on one threesome that seemed to be struggling. 125

TEACHER: What shape are you trying to make?

MOIRA: A diamond.

TEACHER: Can you show me how you would do that?

Each child made and traced a diamond with the pattern blocks; the only difference was one of orientation. They had each used two triangles. 130

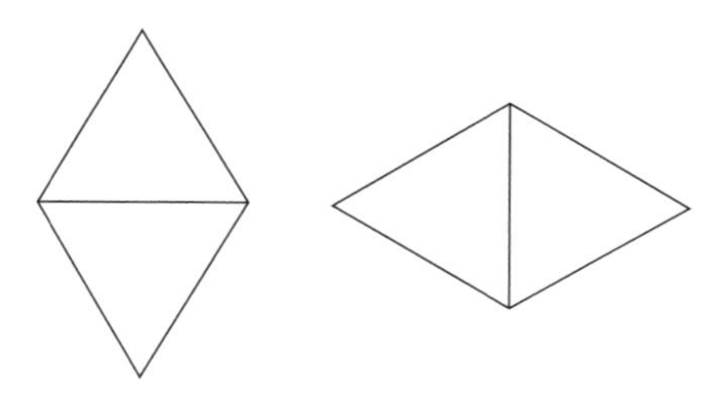

TEACHER: Are those shapes different or the same?

JAMES: They are different.

TEACHER: Can you explain why?

LILY: One shape is bigger and the other is smaller. 135

JAMES: It changes because it gets taller.

MOIRA: They are different because they have different angles.

I was intrigued by this comment and asked Moira to show me what she meant by "different angles."

MOIRA: Well, this shape [the horizontal diamond] has its angles facing a different direction than this shape [the vertical diamond]. The angle of the shape changes, so the shapes are different. 140

I wondered if the students would think the same way about other shapes as well, so I asked this group to build a shape with four triangles and see if their rule would work.

I moved on to check in with Jennifer and Alegra. They were happily creating a number of new shapes and giving them names they had made up. They had two shapes with five triangles that were the same to me, as one was simply turned 90 degrees. Wondering if they, too, were believers in the theory that "different angles make different shapes," I pointed to the two shapes and asked them why those shapes were different. 145

JENNIFER: They are different because one is sideways and one is straight up. 150

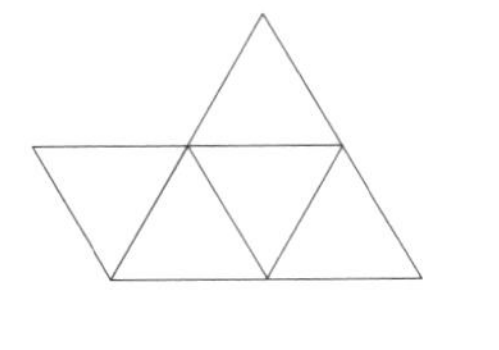
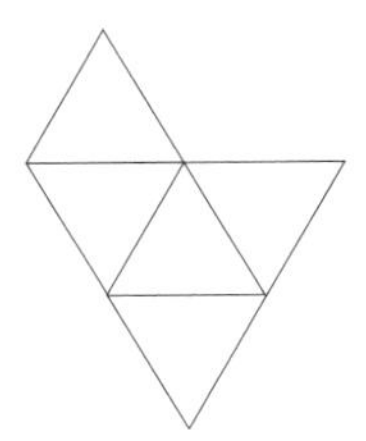

I assumed she thought they were different because of their different orientation, but I wanted to hear how Jennifer would explain her thinking, so I pushed her to tell me more.

TEACHER: What if I took away a triangle from each of these. [*I removed the shaded triangles.*] What shape would we have then? 155

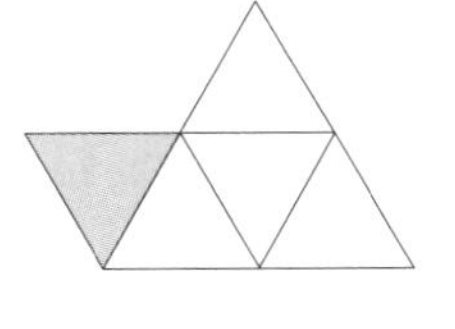
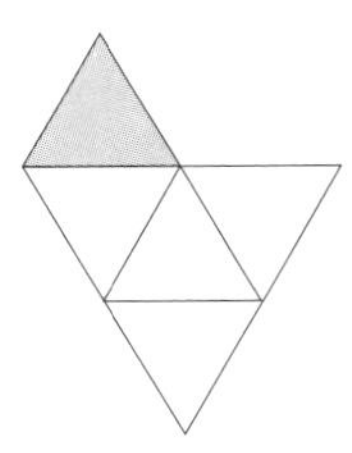

JENNIFER: A triangle.

TEACHER: Would they be the same then?

JENNIFER: I don't know. 160

I wondered why Jennifer had been so sure the two shapes with the five triangles were different, but then was unconvinced when I removed one of the triangles. Exactly how was she seeing the triangles and the shape they made? As the questions ran through my mind, Alegra piped up.

ALEGRA: Yes, they would be different, because if you took off this triangle [*she removes the shaded triangles from each arrangement as shown below*], one shape would be a triangle and the other would be a parallelogram. 165

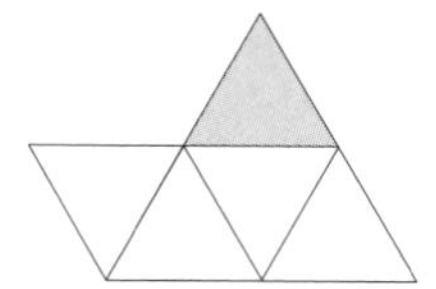
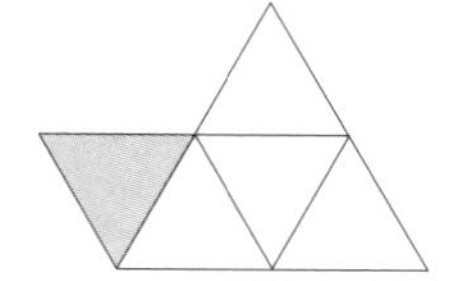

Alegra had made this huge jump, not only considering the orientation of the shape but also looking at removing different triangles to make different shapes. I wasn't sure Jennifer had followed this thinking because she looked puzzled. Although Alegra's thinking was interesting, I didn't want to let go of the orientation idea just yet. I moved them back toward thinking about the triangles turned in opposite directions. 170

TEACHER: Would the shapes be different if they were constructed like these triangles?

175

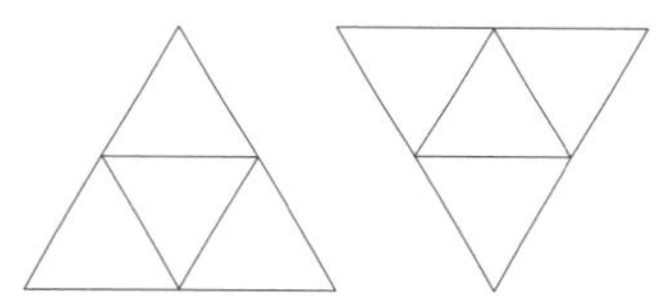

ALEGRA: They are different because of the way the triangles are pointing. One has the tip pointing up, and the other has the tip pointing down.

JENNIFER: I think they are different, too. The triangles would have to be going in the same direction in order to be exactly the same.

It surprised me that both girls were convinced that the two triangles were different, solely on the basis of their orientation. This isn't the way I have been thinking at all, and it made me realize that we had different meanings for the word *same*. They were so sure they had different shapes when the shapes were oriented differently. I was curious to learn how many other children in the class were thinking this way as well. 180

When I called everyone back to the circle to share their work, the first group showed a diamond made from two triangles. 185

TEACHER: Did anyone else make a different shape using two triangles?

A number of hands quickly were raised. One group showed another diamond that was facing a different direction. Holding up the two papers for the class to see, I asked, "Are these two shapes the same?" 190

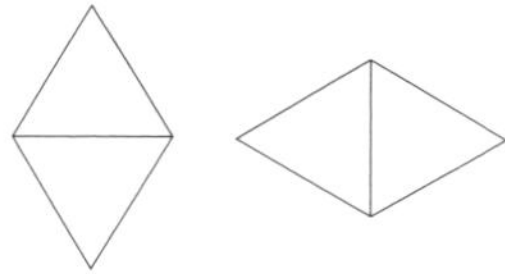

A majority of the class responded no, and their explanations paralleled those I had heard earlier in the small group.

TEACHER: Those of you who said no, why do you think it isn't the same shape?

TODD: One shape is sideways and one is straight up. 195

JAMES: Because this one is a little bit taller than that one.

JULIANA: That shape is going in a different direction.

HARRIET: This one is pointing that way, and that one is pointing up.

As a whole, the class saw the two rhombi as different because they were pointing in different directions. Yet when I had introduced the pattern blocks on the rug the week before, every- 200 one could identify like shapes, and it didn't seem to matter which direction the shapes were facing. I wonder if the fact that the pattern blocks are actual objects, with distinct colors, meant that the children saw these shapes more abstractly? Today when they drew shapes on paper to represent combinations of the triangular pattern blocks, the drawings became fixed on the page in a certain way. Did that make a difference? Did the fact that they were building shapes out of 205 triangles and then looking at the whole shape in a drawing impact the way they thought about the final shape? After all, in most drawings, the position or orientation of a figure does matter: a stick figure standing up is different from a stick figure lying down. However, when I think about the phrase "same shape," I am not paying attention to this distinction. I am beginning to understand how some in the class were thinking. In our next lesson, I'll have them use scissors 210 to cut out the drawings; then we can talk about "same shape" as meaning that one shape can fit exactly on the other. I'm curious to see how that idea will strike them.

case 24

Different or the same?

Sally
GRADE 2, FEBRUARY

In my job as a resource teacher, I have the opportunity to visit a variety of classes. Sometimes I see the same idea popping up in different classrooms. Recently while observing in a first-grade class, I was struck by the fact that the students believed that two congruent rectangles oriented differently were not the same. The students were looking at two rectangles: one oriented vertically (standing on its shorter side), the other horizontally (with its longer side as the base).

Even though the rectangles had the same dimensions, the students did not consider them to be the same. As they explained their thinking, I realized that for them a rectangle is long and skinny, a "pulled out" square. They argued that a square could be pulled out from side to side to make a rectangle, but they did not accept that it could be stretched from top to bottom. They concluded that a tall, narrow box—a simple rotation of the long and skinny box—was, in fact, not a rectangle.

Later as I entered a second-grade class, I found myself face-to-face with these very same rectangles! One was on its side, long and skinny; the other was standing up, tall and narrow. Both rectangles, as drawn, were divided into six tile-like squares.

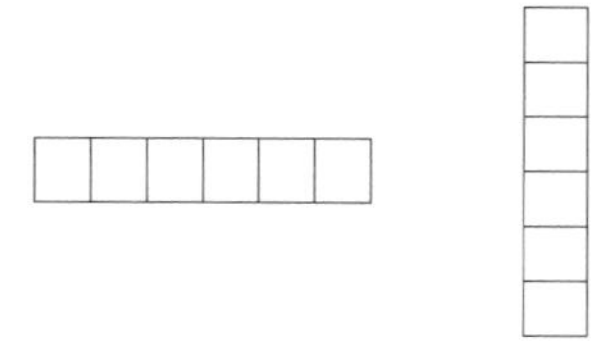

The student teacher stood at the whiteboard and asked, "Are these two rectangles the same or different?" As students offered opinions, the teacher recorded the votes with tally marks below the labels *Same* and *Different*.

CESAR: Same, because they both have 6 tiles.

JILL: Same, because they both are rectangles.

LEXI: Different.

FRED:	Same, because one covers the other.	235
MICKEY:	Different.	

According to the tally marks, twelve students thought the rectangles were the same; the other eight felt they were different. From a pre-class discussion with the teacher, I knew that she wanted to use this question to lay some groundwork for an investigation into the idea of congruence. I thought about the students' responses and wondered about the eight votes for "different." Were "same" and "different" adequate choices for focusing the discussion? Did the rectangles feel different to the students for reasons that we don't know? (Only those who said "same" gave a reason.) Did those eight students share the perception I noted in the first-grade class, that a different orientation makes the rectangles different? In the first-grade class, the rectangles were drawn like the outline of a box; here they were divided into six squares. Did that change what the students were paying attention to?

As the lesson continued, the students reviewed yesterday's classwork, "Using six tiles, make all the different rectangles you can." They had decided they could build just two different rectangles, a 3 ×2 and a 6 × 1. The teacher drew these two different rectangles on the board.

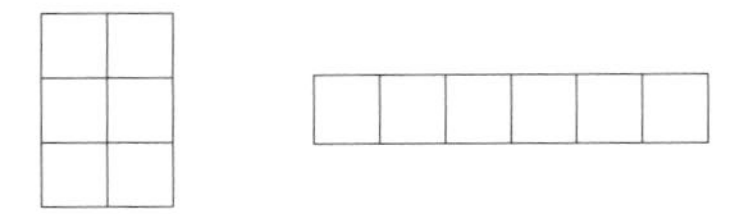

I wondered about the eight students who had voted that today's rectangles, the one standing and the one lying down, were "different." Employing the criteria that they used for their vote, wouldn't those same students now be thinking that there should be four drawings, for four different rectangles: the 3 × 2, the 6 × 1, and the two "different" ones, a 2 × 3 and a 1 × 6?

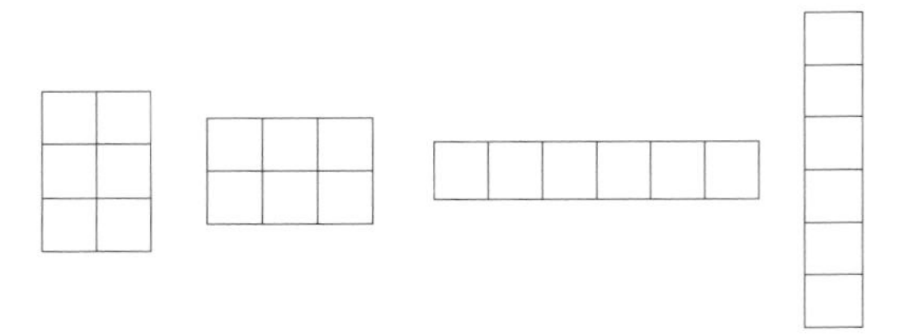

Wouldn't that be consistent? Why had none of those eight students questioned the "class" conclusion? I know that I am haunted by it.

After the warm-up and review, the teacher gave the instructions for today's classwork. Armed with a baggie of 12 one-inch square tiles and 8 × 8 grid paper, the students set to work independently to find how many different rectangles they could build.

What would happen to the student thinking? Would building the rectangles with inch-square tiles and then drawing them on grid paper offer the students new insights? Would the investigation offer them ways to extend their thinking about the attributes of rectangles?

I joined a group of three students who got to work pretty quickly.

240 245 250 255 260 265

LEXI: Look [*She counts by fours with confidence*]: 4, 8, 12. Three rows of 4 is 12. Just add 4, three times! [*She went on to build her next rectangle, 6 rows of 2.*] Look: 2, 4, 6, 8, 10, 12. Just count by twos.

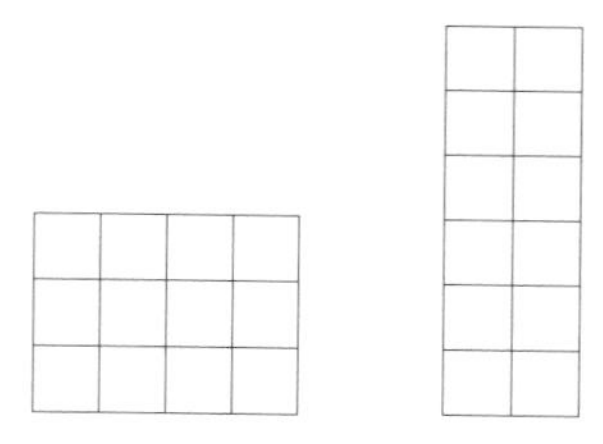

In the same group, Mickey had used all 12 tiles, but placed them in an L shape, 8 down and 4 more across. 270

MICKEY: [*perplexed*] Could this be right? I know this isn't it, but I can't do it.

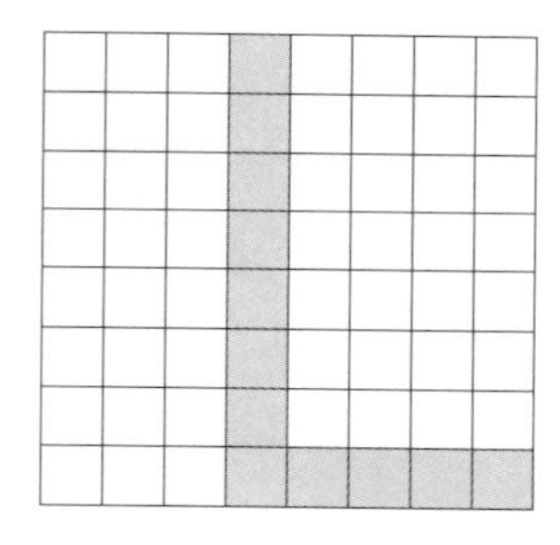

I'm not sure what he is confused about. I asked him what the assignment is. He told me it was to make rectangles with the 12 tiles. While I was thinking about where to go with Mickey's confusion, Toby, in similar frustration, almost leapt across the desk. 275

TOBY: Hey, is this a rectangle?

I was surprised that both boys had this question because I knew that this was their second full week of rectangle work. I looked across at Toby's paper and saw an upside-down T, 8 tiles down and 5 across the bottom. 280

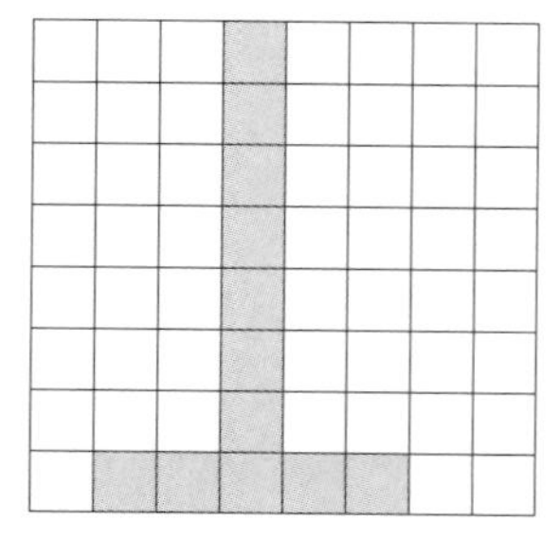

MICKEY: I know it can't look like this [*referring to his L shape*] and be a rectangle. I know it has to go straight down.

It is suddenly clear to me what has happened: both boys, in attempting to build a 12×1 rectangle, had been confused by the limitations of the 8×8 grid! 285

Even though Lexi was in this group, she continued working confidently by herself. She cut out her two rectangles, the 6×2 and the 3×4, and glued them onto construction paper. Watching her, Toby got an idea. He cut out the T shape and then cut it into two parts, a 7×1 and a 5×1. He rotated one and slid them together to make a 12×1 figure. 290

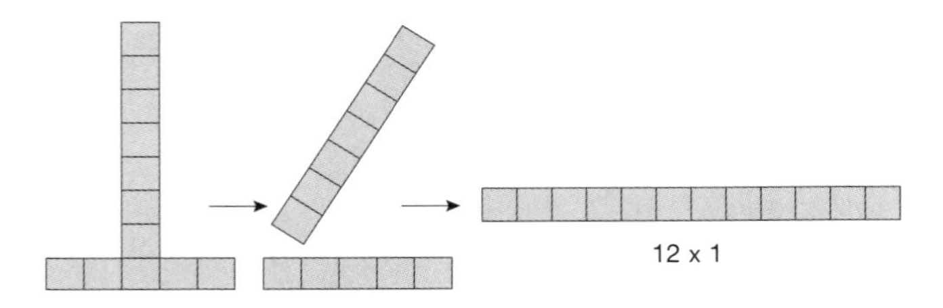

TOBY: This [12×1] is a rectangle.

Meanwhile Mickey had given up his L-shape dilemma and was now fascinated by the fact that his 2×6 cutout matched the 2×6 hole next to it on the paper. He rotated his cutout so the two rectangles (the hole and the cutout) looked like the two on the board this morning. 295

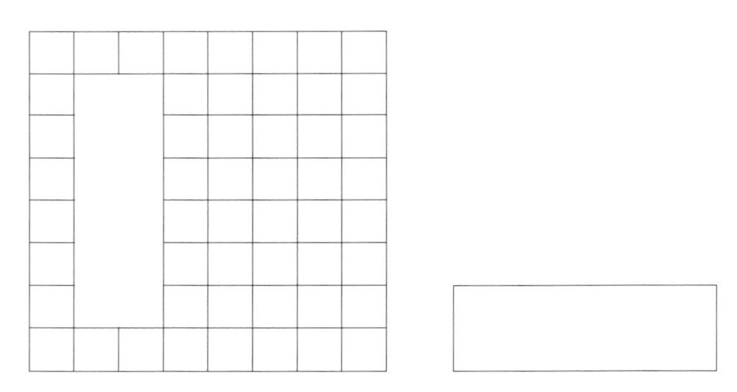

MICKEY: Are these the same or are they different? I don't know. I voted "different" before, but now . . . [*he picks up his cutout and fits it back into the hole*] I think I can prove they are the same.

Wow! What next? I do want to find a way to bring Toby's and Mickey's new ideas to the whole class, and I know I'll do that. However, I am also left wondering: What was it about this work that helped them to see that a 2×6 rectangle is still a 2×6 rectangle, no matter how it is drawn? We use the word *same* in English to mean different things. In some ways, all rectangles are the "same," but here I want them to see the meaning of *same* as the geometric term *congruent*, that is, that one figure can be turned to fit exactly on the other. Now that I have seen what "same shape" means to most of the students, I'll have to think about how to help them take in the meaning of *congruent* as it applies to geometry. Mickey's "proof" will start the discussion. 300 305

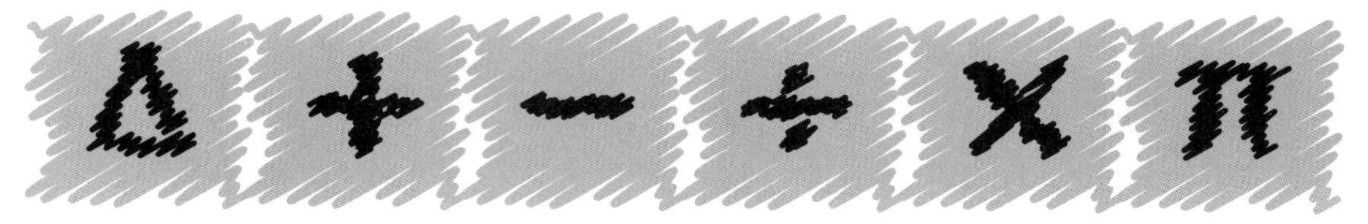

CHAPTER

6

2-D images of a 3-D world

CASE 25
What's 2-D and what's 3-D?
Eva
Kindergarten, January

CASE 26
What about the
bird's-eye view?
Ellie
Grade 1, April

CASE 27
The math in a chair
Natalie
Grades 3 and 4, February

CASE 28
Describing geometric solids
Olivia
Grade 2, November

CASE 29
Boxes
Anita
Grades 3 and 4,
February–April

The first chapter of this casebook included children's comments as they described 3-D objects. Some of those descriptions acknowledged the 3-D aspect of an object and some did not. For instance, some children described a cube as a "fat" square, others as just a square. In still other situations, children have noted that a cube is made up of six squares (they are noticing the six faces). This range of children's descriptions of a cube leads to general questions about the relationships between 2-D and 3-D objects and the ways children think about them. In what ways do children notice the "3-Dness" of an object? How do they make sense of the way flat surfaces fit together to make a solid shape?

In this chapter, we revisit these questions and extend them to look at the way children draw 3-D objects.

- What do children take into account in their drawings?
- How do they represent 3-D relationships in 2-D drawings?
- How do they form an image of a 3-D object from a 2-D representation?

As you read the cases, consider your answers to these questions and try to identify the issues that children must work through in order to make sense of these relationships.

case 25

What's 2-D and what's 3-D?

Eva
KINDERGARTEN, JANUARY

Today's activity for exploring geometry was a "shape hunt." I handed out a worksheet with eight 3-D shapes pictured on it, and my kindergartners had to look around the classroom to find objects representative of these shapes.

Many of the items they found were good examples. For instance, they chose a tin can holding pencils for a cylinder, a jack-in-the-box for a cube, and a blackboard eraser for a rectangular prism. However, several children found 2-D shapes and claimed they also matched the 3-D shapes on the sheet. One girl found a painted picture of a house; she outlined the three sides of the triangular roof with her finger and said it matched the triangular prism. She may have been thinking of an actual roof rather than the triangle drawn on the paper to represent it. However, two boys found circles and squares cut from construction paper and chose those to represent the sphere and the cube. 5 10

Teams of children were grouping 2-D and 3-D shapes and objects together and matching both kinds of objects to the 3-D shapes depicted on the sheet. I was curious about their decisions. What were they paying attention to as they matched shapes? I approached four children who were working together to find out what they were thinking. I began our conversation with a question about two wooden blocks—triangular prisms of different sizes and shapes—that I held in my hands. 15

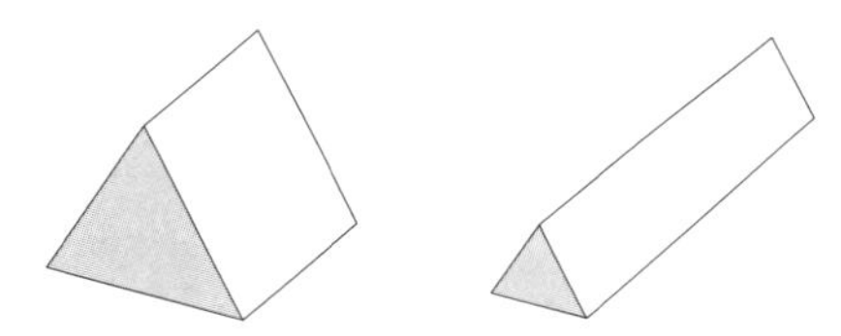

TEACHER: Are these the same? Do they belong in the same group?

JOHN: Yes, one is bigger and the other is smaller. 20

IRENE: One is just skinnier than the other. This side is longer too.

TEACHER: [*holding up a paper triangle and a thin plastic triangle*] Do these belong in the group with the two wooden pieces?

CARMEN: They all go together because they are triangles.

TEACHER: Do they belong with these [the two wooden triangular prisms]? 25

CARMEN: Yes, they have triangles here. [*She points to the triangular faces.*] They are the same; these [the wooden blocks] are just longer.

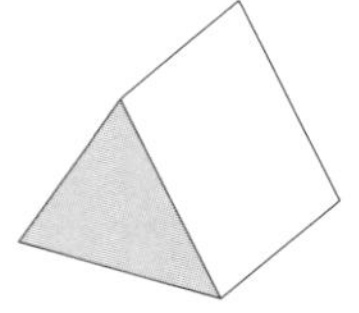
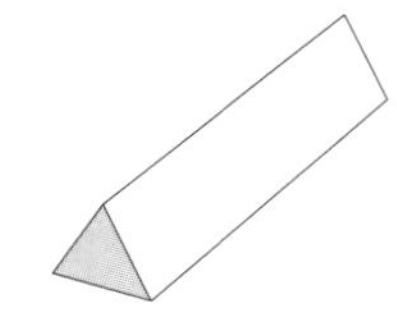

Carmen's statement makes me think that she does recognize the third dimension. She refers to it by saying the shapes are the same, but then points out a difference: The prisms are "longer" than the paper triangle. She sees the third dimension, but doesn't yet have the correct words to describe it. I do wonder, though, why she doesn't think the added dimension is different enough to warrant a separate category. Exactly what about the shapes is she focusing on? Maybe the thinness of the plastic and paper triangles isn't as important to her as the "triangle-ness" of all four objects. 30 35

TEACHER: [*holding a tennis ball, marble, and a wooden sphere*] Why did you put these in the same group?

IRENE: They're all circles, so they go in the same group.

TEACHER: [*holding up a circle drawn on a piece of paper*] What about this one? 40

CARMEN: That goes with the others, because it's a circle, too.

TEACHER: They look different. Why do they go in the same group?

CARMEN: One is flat and the others are higher, but they're all circles.

Again, Carmen says they are the same, but then indicates the 3-D shapes are "higher" than the paper circle. She sees the third dimension and uses the word higher to denote it. I think she wants to say they are different, but doesn't know what else to call them, so she says they are circles. I know young children have lots of experiences at home and in preschool recognizing the basic 2-D shapes, so it is not surprising that Carmen relies on these shape names to describe the 3-D shapes too. I think she is ready for more activities centered on the third dimension. 45

Holding two wooden cubes of differing sizes, I ask the children if they belong in the same group. 50

JOHN: Yes, they go together; one is bigger and one is smaller.

TEACHER: What do you think, Kate?

KATE: I think they belong together because they both have squares. [*She points to a square face on each cube.*] 55

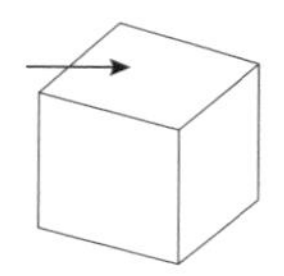
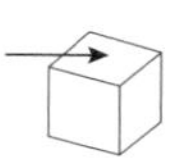

TEACHER:	[*holding up a square drawn on a paper*] Does this belong with them?
KATE:	Yes, it has a square too. They all have the lines: 1, 2, 3, 4 [*As she counts, she points to each line of the square. Then she points to the top of one cube and counts the four edges.*] 1, 2, 3, 4. They are all the same. They all have four lines. 60

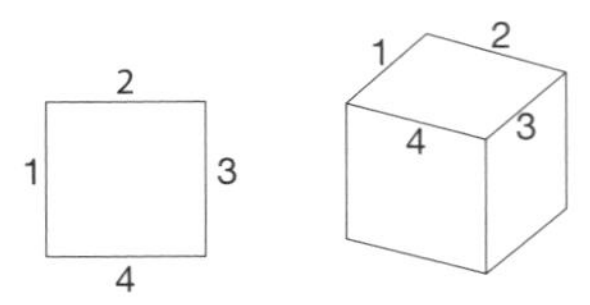

I see that Kate notices the similarities between the square and the cube, but not the obvious differences. She doesn't appear to discriminate between a line and an edge. I wonder if she notices that there are other "lines" on that cube besides the ones she counts? 65

CARMEN:	[*agreeing with Kate*] They are the same.
TEACHER:	Don't they look different?
CARMEN:	Yes, but if you have more of these [squares], then you could make a block.
TEACHER:	What do you mean, "if you have more of these?"
CARMEN:	If you have more papers, you can put them all together and you can have a block. 70
TEACHER.	Do you mean to pile them up one on top of the other?
CARMEN:	Yes, it is too flat. If you put lots of the papers together, you have a block.

I think Carmen consistently discriminates the dimension of depth. I like the way she notices that the block is solid. She isn't just using six squares to build the sides of the block, but is accounting for its thickness. John interrupts my thinking with his comment. 75

JOHN:	They belong in the same group, but this one [the cube] is different [from the square]. It has a line here [*pointing to a vertical edge on the cube*]. The other one doesn't have a line. 80

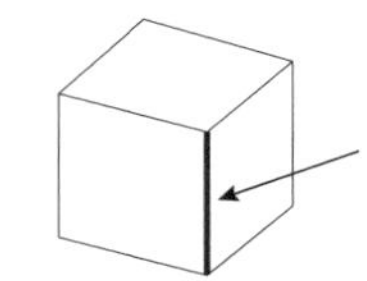

IRENE:	They are all squares; one is just flat, that's all.

So John acknowledges that the cube has a property that is different from the square, but he doesn't know what it is called. He relies on the familiar word of *line*. Like Carmen, he is beginning to discriminate between 2-D and 3-D shapes and is focusing on some of the important differences. Irene is paying attention to one aspect of shape, the squares. While she notes the depth 85 of the cube, it doesn't seem to be an important element to her. I think I need to provide some experiences that will take them one step further and stretch their thinking about this dimension they are now beginning to recognize.

To take stock of where these kindergartners are in their thinking: They are able to recognize 3-D shapes and common representations of them in their environment. They are formulating 90 their thoughts about the relationship of these shapes to the more familiar 2-D shapes (triangle, circle, and square). They often group 2-D and 3-D shapes together and say they are the "same" on the basis of common forms contained in the 3-D shapes. When asked about the differences, children may note the third dimension, but don't seem to think it is significant. I think they regard this difference as some type of minor variance in shape. Irene is at a stage in her devel- 95 opment where she is wholly unconcerned about the differences: "One is just flat, that's all." Her interest in the differences has not been aroused by this activity. In contrast, Carmen's and John's expanding knowledge of the 3-D shapes is clearly evidenced as they begin to separate them from their 2-D counterparts.

case 26

What about the bird's-eye view?

Ellie

GRADE 1, APRIL

100

While working through our first-grade geometry unit, we recently began to experiment with drawing 3-D structures. Today's lesson asked the children to build a structure using some wooden blocks in geometric shapes, to draw a picture of their structure, and then to give their drawing to a partner. The partner was to build the same structure, using the drawing as a building plan. 105

It was fascinating to see the children grapple with representing their structures on paper. At each table, children experimented to find the best approach. There was a lot of conversation. Jennifer, Todd, Emma, and Neville decided that it was really important to draw the structure face on and not to bother with the sides they couldn't see. At another table, the children decided the easiest approach would be to trace the shapes and then add lines to show three- 110 dimensionality. A third group decided to draw from a bird's-eye view, drawing what they would see looking down on the structure. The class was clearly engaged, and I was really looking forward to seeing how the children would build from each other 's drawings.

I collected the finished drawings and redistributed them with the instruction that children should try to replicate each structure as drawn. Within minutes, tension could be felt in the 115 classroom.

> "I can't find the block I need!"
>
> "I don't understand this drawing!"
>
> "I need help!"

I let them struggle for a while before I called them together to discuss what had made the 120 task so challenging. I began by asking if anyone had been successful with the building task. It turned out that only one child, Bo, had been able to re-create the structure from the drawing he was given. The other children were interested in what he had done and what the drawing looked like. Bo told us that Elsa had made the drawing, and he displayed it for the class.

I asked the children why the task was easier for Bo than it had been for them. 125

Fig. 6.1. Using this drawing as a plan, Bo was able to replicate the original structure.

ALEGRA:	It's easy to tell the blocks Elsa used.	
TEACHER:	What makes it easy to tell?	
ALEGRA:	She made the sizes almost the same as the real blocks.	
TEACHER:	Anything else that helps?	
BO:	Elsa drew all the blocks. There were none that were hiding.	130
JAMES:	She made it look three-dimensional, so we could see the sides.	
MOIRA:	She didn't make a bird's-eye view.	
TEACHER:	Why does drawing from a bird's-eye view make the building job more or less challenging?	
ANDRE:	Because you can't really see how all the blocks are standing.	135
JULIANA:	Plus it's hard to tell the difference between the shapes when they are drawn on top of each other. You can't see if it's a triangular block or a square block.	
TEACHER:	Were there any other drawing methods people used that made it hard to see the shapes?	
JAMES:	Yes, it's also hard when you trace the shapes because they don't come out looking like the structure.	140
TEACHER:	Did anyone else who tried to trace the shapes find it challenging?	
NONA:	I started to trace them and gave up because it didn't look right.	
TEACHER:	What didn't look right?	
NONA:	I traced the bottom of the block [a triangular prism] and then I couldn't figure out how to make the triangle sides.	145

At this point it was time to head to music. I told the class we would do the same activity tomorrow, and asked them to remember what we had discussed. I had decided that the children needed to do this activity a second time. I was curious to see if they would remember the reasons their classmates had given for what made it so difficult to replicate the structures, and if they would then do their drawings differently.

As we began again the next day, we reviewed some of the ideas from the day before. The children stated clearly that they would not trace the shapes or draw from a bird's-eye view, since both ways made it impossible to figure out the shapes that were used. Even the children who had originally been convinced about tracing or drawing from a bird's-eye view were now agreeing with this. Once the children began working, I went to talk to some of them individually, starting with James, who had traced his blocks for yesterday's drawing.

TEACHER: Why did you decide not to trace the shapes this time?

JAMES: Because it was too hard to draw the sides after I traced the shape. People had a hard time telling which shapes I used to build my structure.

Only yesterday, James was convinced that tracing was the best method. His change of heart was encouraging, but James is a child who struggles with spatial relations and I was curious about how he would draw his structure this time. I moved on to Harriet, who had used the bird's-eye view approach yesterday. When I got to her, she had just finished building her structure and was ready to begin drawing.

TEACHER: How have you decided to draw your building?

HARRIET: I'm just going to draw all the shapes.

I stopped next to talk with Todd, who previously had difficulty representing the size of the different blocks. The class had mentioned that it was easier to re-create a structure if the blocks were drawn to scale. I could clearly see in his second attempt that he was trying to reflect the size of some blocks. However, he continued to represent all the blocks much smaller than they actually were, without taking into account that some blocks were much larger than others. Thus, his drawing didn't fully convey to the builder all the different sizes he had used. He did make adjustments for the tiniest blocks, so that was a beginning.

As I continued around the room, I was impressed with how much of the previous day's conversation the children had taken into account as they drew their structures for the second time. Even though their drawings were of different structures from one day to the next, I could see progress in how they were taking on this work.

180

Fig. 6.2. Todd's second drawing is clearly an improvement over the first day's tracing.

Fig. 6.3. Harriet gave up the bird's-eye view approach she had used the first day in favor of "just drawing" the blocks.

I was glad I had decided to have the children repeat the block drawing. Clearly the discussion after the first day's efforts had given the children in the class a reference point for their drawing on the second day. James, who traced his shapes and came out with an indecipherable picture on the first day, made a significant change in his approach on the second, producing a drawing a friend could read and know what to build. Harriet was now aware that she needed to include all the shapes so someone else could know which blocks they needed. And with his second drawing, Todd tried to represent size differences to help a builder choose the right blocks. 185

Had I not given them a second chance, would they have been able to recognize the changes that needed to be made? This highlights the importance of allowing children the time to think through a process and the time to try it again. Often the pressure of time stands in the way of building on an idea. I was glad I had resisted that pressure and allowed for a second day of building and drawing. 190

case 27

The math in a chair

Natalie

GRADES 3 AND 4, FEBRUARY

Art is an integral part of the learning that takes place in my classroom; I integrate art into many subject areas. Although it isn't always clear to me what links the two disciplines, I am particularly intrigued by the connection of art and mathematics. I've always had a vague sense that geometry is about shapes, and I know that in drawing, it's important to recognize the relationship between shapes. But only after listening closely to my students in one series of art lessons did I see how much math was embedded in our artwork. 195

For one art assignment, the students were drawing from real life. As I circulated around the classroom, I noted that Nick, who was trying to draw a chair, was visibly frustrated. I sat down and reviewed his work with him. I decided to be explicit about ways to improve his sketch. Together we looked at his drawing and also at the chair itself. I told him what I noticed looking at the chair, pointing out the lines and angles that I saw. As I talked, I realized that my language sounded more mathematical than it did artistic. I said things like, "Notice the parallel lines here," or "When you draw the angle of the bars . . . ," or "It's like a cube." I left him for a while and when I came back, he was clearly pleased with his revised drawing. 200 205

Fig. 6.4. On the left is Nick's work before our conversation; on the right is the chair he drew after we talked.

I was struck by the difference in his pictures and became intrigued that the mathematical terms had been helpful. Before we had talked, he was frustrated because his drawing made it 210 look as if the supports under the seat crossed—and he knew they didn't touch. When I thought about the language, it seemed to me that I had stripped away the scary, mysterious part of art—how to draw a chair realistically—by showing Nick that the chair was somewhat like a cube. He knew how to draw a cube and show its three-dimensionality, something he wasn't able to do when drawing the chair. Applying what he knew about cubes allowed Nick to get beyond his 215 mental image of "chair" and use what he already knew about drawing in three dimensions.

As a result of Nick's work, I decided to have everyone draw a chair. I wanted to pursue some of the mathematical ideas, even though I wasn't sure exactly what those ideas were. The next morning I placed a chair in the middle of the meeting area with the message, "Draw a picture of a chair." I then watched as my students made their sketches. Later, we talked about what 220 they noticed in one another's drawings. They began by commenting on how the pictures were drawn, such as the quality of the pencil marks and the amount of detail, but then Nate said he saw "a couple of pictures that were three-dimensional."

TEACHER: What do you mean by three-dimensional? 225

LYNDA: I think it means that you are drawing what you see—not something flat. You are drawing something that looks like you could touch it, like it is coming off the page. The chair isn't squished.

JODIE: Two-dimensional is the width and height. So if it were two dimensions, it would have width and height, but it wouldn't be popping out of the page. For 230 example, Michael's is flat; it's not popping out.

TEACHER: So how do you make it three-dimensional?

NICK: It might be the angle. You could draw the different pieces coming out.

HUNTER: In three dimensions, you wouldn't draw everything. You would only draw what you see. If you are looking at the chair from a particular direction, you 235 only see pieces of it.

I was struck by this last comment, that in 3-D drawings, "you wouldn't draw everything." I take this to mean that if you look at a chair, you may know that there are slats between the legs, four legs, a back, and a seat, but that doesn't mean you would draw each of those pieces. Drawings by young children often look as though they were drawn from above, but they also try to 240 show all the pieces, even pieces that couldn't be seen from that position. Hunter was saying that while he knows there are separate shapes that make up the whole shape, you might not draw them all.

Intrigued by the issues and the problems involved, my students have continued to draw chairs and talk about their work. These have been engaging conversations about the mathe- 245 matics they've discovered in each other 's artwork. The focus of the discussions has often been angles, or what they sometimes call "slanted lines." They have become increasingly aware of the fact that changing the direction of a line can make something look three-dimensional or more realistic.

When I look at their pictures and hear their conversations, they seem to be struggling with how to move shapes around in space. They know the chair is made up of shapes, specifically rectangles and squares. But what they know and what they see are sometimes two very different things. So I am thinking of these math questions: How do children see, or understand, the relationship between two or more shapes within a bigger shape? How do they maintain the three-dimensional relationships when they have to convert the solid object into a two-dimensional drawing? For example, while the seat on the actual chair is a square, in the drawing it could be a trapezoid or a parallelogram, depending on the viewpoint. What mathematics do children work through that allows them to adjust the shape they know to the shape they presumably see? What do children need to do to coordinate the elements of an object so they fit together?

Some of my children easily render three-dimensional objects as drawings that appear to be 3-D. They recognize and account for spatial relationships. In their work, squares become reconfigured into trapezoids, circles into ovals, and so forth. These students know every element of a chair, but they also know not to draw every element. Both Pushpa's and Susan's drawings offer examples (see fig. 6.5). In Pushpa's drawing, the seat of the chair is a square, but she didn't draw it that way. I also see that she didn't draw in the second front leg of the chair, because it is not visible from her point of view. For the chair that Susan drew, even though in reality there is a space between the legs and the supporting rungs, she didn't insist on that space in her drawing.

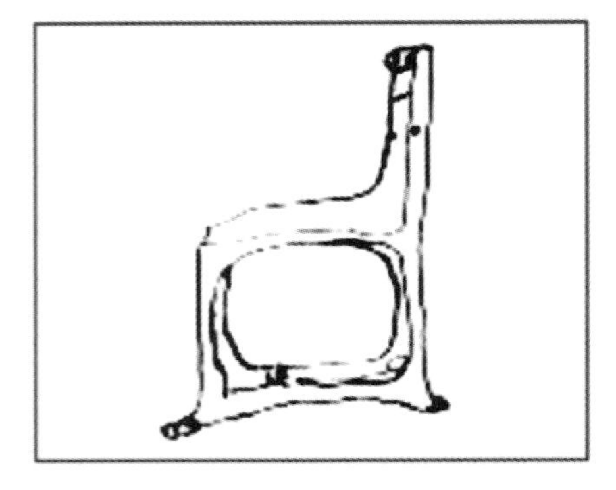

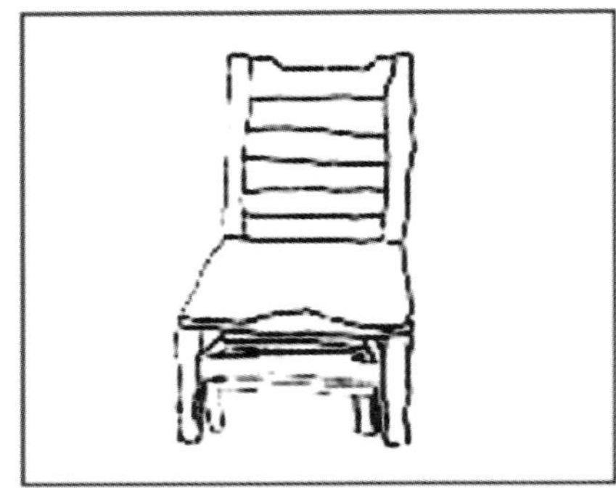

Fig. 6.5. Pushpa (left) and Susan (right) are starting to recognize spatial relationships as they draw 3-D chairs.

While some children's drawings represented the chair realistically, others did not. Some drawings looked contorted or out of whack. In the past I would have chalked this up as typical for this age. Now I look at the drawings with a new eye. What do these chairs tell me about the way a child can or cannot manipulate shapes in his or her mind and then reinterpret them on paper? Rotating, changing, and coordinating the surfaces and angles in their heads to make realistic representations takes a particular kind of mathematical agility.

Fig. 6.6. Jake is struggling with 3-D representations.

In his two efforts in figure 6.6, Jake has drawn every element, whether or not each would be visible from a given perspective. In both drawings, the point of view is distorted. No place in the room would offer such a view, not even if you were looking down from the ceiling. These  280 drawings have a cubist or a Picasso feel to them. In the first drawing, the four different sides of the chair (the back, the seat, the front leg, and the side legs) are all shown from a different point of view. In fact, the back is not even connected to the seat, showing me that there is real struggle here for this student when coordinating the relationship between the two shapes. His drawing reminds me of a cube opening up. 285

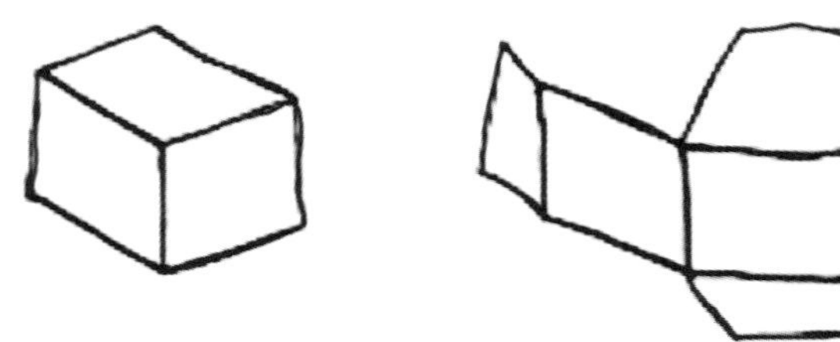

Other drawings reveal the difficulty some students have with representing the way the lines in the chair intersect. In reality, chairs are usually constructed with legs that are perpendicular to the seat and seats that are rectangles. In a drawing, however, some of the right angles might have to become acute or obtuse angles in order to show the 3-D depth on a 2-D sheet of paper. 290 For instance, seen head-on, the seat of a chair might be drawn as a parallelogram with two acute and two obtuse angles, rather than as a rectangle with four right angles. Some children find it hard to give up what they "know" in reality about a chair and persist in drawing right angles.

Lynda (see fig. 6.7) is trying to figure out the differences between drawing a 2-D chair and a 3-D chair. She has some notion that to represent a chair more realistically means she needs to 295 draw more detail. I see she has flipped the chair around to show the connecting bar in her 3-D attempt. I also notice the angles are all still mostly 90°. Her image of the chair as an object containing 90° angles is so strong that Lynda isn't able to draw the chair without those perpendicular lines.

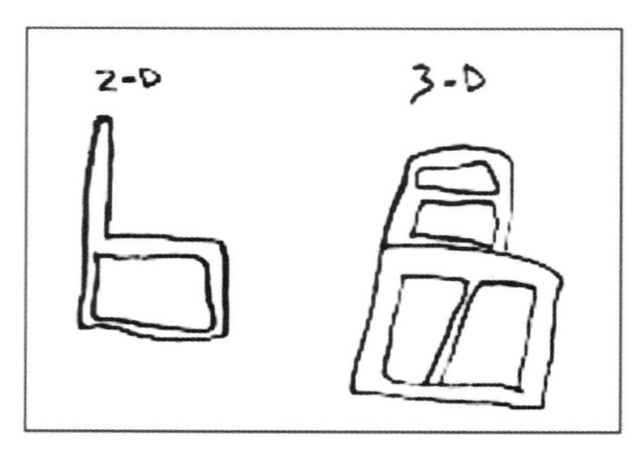

Fig. 6.7. Lynda has been unable to give up the idea that a chair is constructed with 90° angles.

300

And finally there is Bill's drawing (see fig. 6.8). When he first showed this to me, I didn't know what to make of it. I couldn't even see the chair in the drawing. I knew he had been working on drawing an overstuffed chair that sits in our reading corner.

BILL: I'm having a hard time forming this chair. I just drew what I saw—it's a bunch of shapes. I just drew them together. You can't draw in 3-D. 305

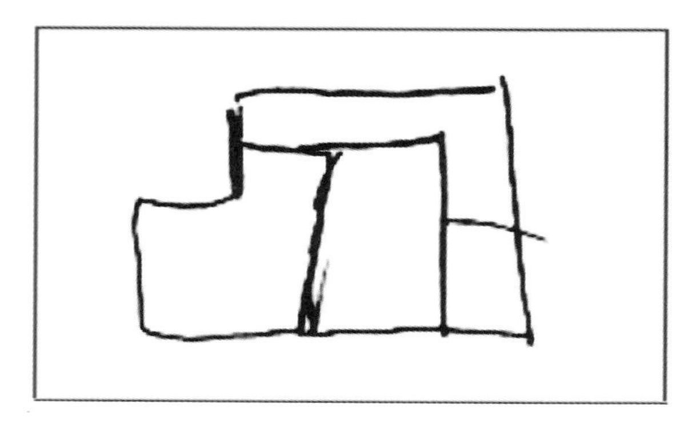

Fig. 6.8. Bill can see the shapes that make up a chair, but he is unable to connect them realistically.

I still have a hard time seeing the chair in his sketch, but now I can guess what was going on for him. It looks to me as if Bill saw that the chair was made up of squares and rectangles, but he couldn't figure out how those shapes connected to one another. Bill's picture may be extreme, but it makes me aware of the amount of work children need to go through to transform shapes in their minds. 310

This chair-drawing exercise has given me a better grasp of the links between math and drawing. So much of mathematics is about looking for relationships and the ability to take apart elements and put them back together to construct meaning. Going back to the case that started 315 me off, Nick built on his prior knowledge of a cube to help him draw a chair. He separated out the different aspects of the assignment—the artistic image of the chair, the actual parts of the chair (its legs, back, and seat, as well as the invisible sides), and his mental image of a chair. Piece by piece, he reconstructed the shapes to fit together into a realistic portrait. And that part of the assignment was all mathematics. 320

case 28

Describing geometric solids

Olivia

GRADE 2, NOVEMBER

In my job as math resource teacher, I have the opportunity to work in a number of different classrooms with students of different ages and at various developmental levels. For the past few weeks, I've been working with a group of bilingual second graders. I love my time with these students. Not only am I able to use my Spanish and my skills as an ESL teacher but I am also 325
bowled over by their enthusiasm for the math, too.

In the geometry sessions to this point, the second graders have explored both 2-D and 3-D objects. They have used plastic polygons to fill in the outlines of larger shapes, worked on finding the footprints of 3-D blocks, played games based on sorting objects by attributes, and generated lists about the number and kinds of faces on different 3-D blocks. 330

One activity we did was Find the Block, (from Attributes of Shapes and Parts of a Whole, a second grade unit from *Investigations in Number Data and Space,* 3rd ed., Pearson 2016) in which students locate a particular geometric solid by examining a set of drawings depicting all its faces. Thus, a card with six squares would represent a cube, a card with one square and four triangles would represent a pyramid, and so forth. 335

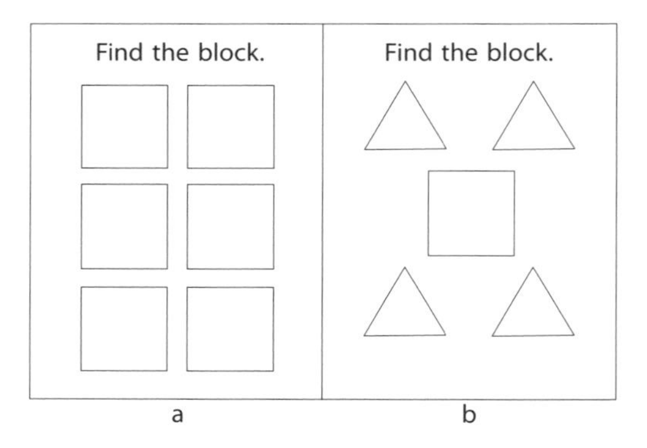

Figs. 6.9a. and b. Find the block.

It struck me that Find the Block really pushed students to look at each block in a particular way, focusing their attention on the number and shape of its faces. I wondered how the students would respond to a more open-ended activity, such as describing the blocks in writing. At the beginning of the next class, I gave each student a single block and unlined paper. 340

TEACHER: Today I'm going to ask you to do two things: I want you each to look very carefully at your block. On the top half of your paper, I want you to draw a picture of your block. It doesn't have to be perfect. Just do the best you can.

> After you've drawn your block, I want you to write on the bottom half of your paper at least three things that would describe your block. You can write more than three things, but you can't write less. 345

I wanted them to draw the blocks for two reasons. First, I thought it would help them to focus their attention both on the characteristics of the blocks and on the whole block itself. In addition, I was interested in seeing whether and how they would attempt to show that the blocks were 3-D. The results were fascinating. (I have provided my own translation when they wrote in Spanish.) 350

I had given Zulma a cube. I found it interesting that her drawing (see fig. 6.10) showed only one square face, probably traced. She had initially tried to figure out how to show the 3-D aspect of the block, become frustrated, erased her original drawing, and settled on showing a single square face. Her statements—that the block has 6 faces, 6 squares, and looks like a diamond when you turn it—showed me that she was thinking of all of the faces. 355

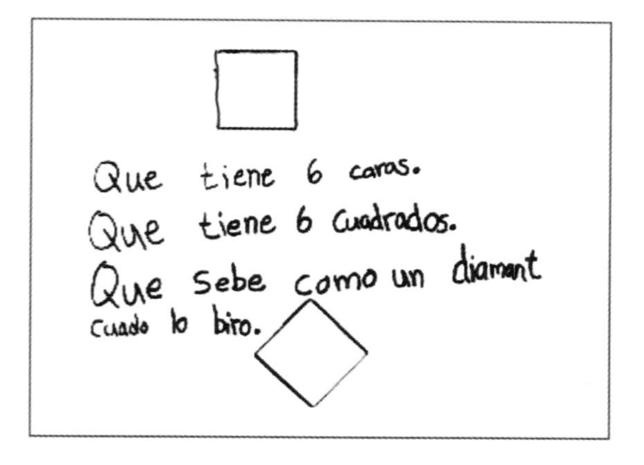

Fig. 6.10. Zulma's description of a cube

Zulma then chose a second block to describe: a rectangular solid with four rectangular and two square faces. This time her drawing showed the outlines of one square and one rectangular face (see fig. 6.11). Her statements included one about the number of points on her block. She wrote that this block has 2 squares, 6 faces, and 8 little points. 360

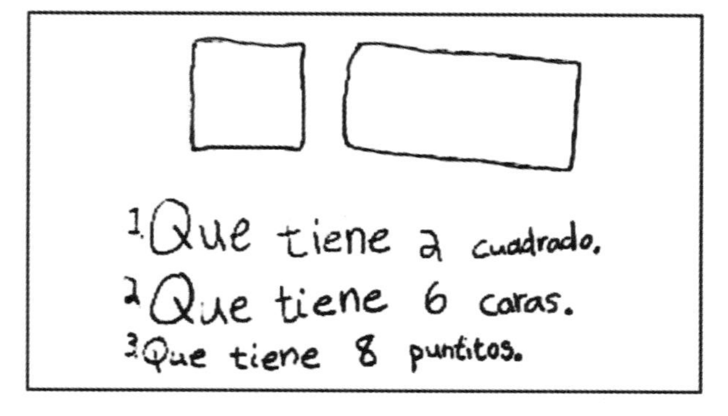

Fig. 6.11. Zulma's description of a rectangular prism

Paco's drawing (see fig. 6.12) was similar to Zulma's approach. He had been given a triangular prism, which he described as having 1 cuadrado (square), 2 rectángulo (rectangles), and 2 triángulo (triangles). Paco, like Zulma, had chosen to draw his block by representing only one of each of the identified faces. 365

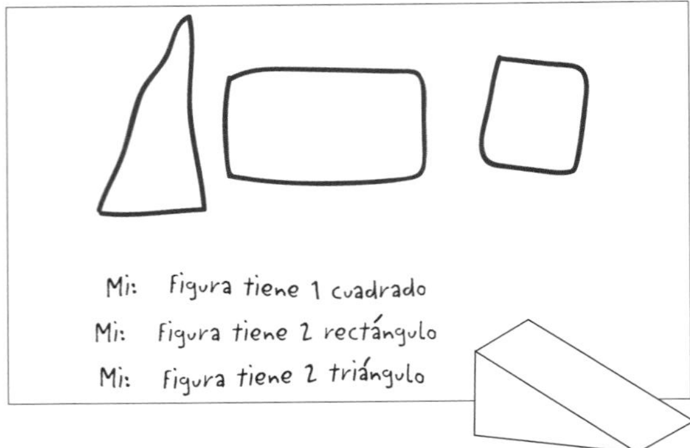

Fig. 6.12. Paco's description of this pictured triangular prism

Brenda was working with a cube. Like Zulma, she had struggled with her drawing, but she resolved the difficulty by representing all six faces (see fig. 6.13). I was intrigued by the fact that the faces in her drawing all touch one another; perhaps this was her way of trying to show their connectedness. She wrote, "I like it because it is a square and it has 6 faces and it is big and it has 8 corners and it is pretty and I like the faces." 370

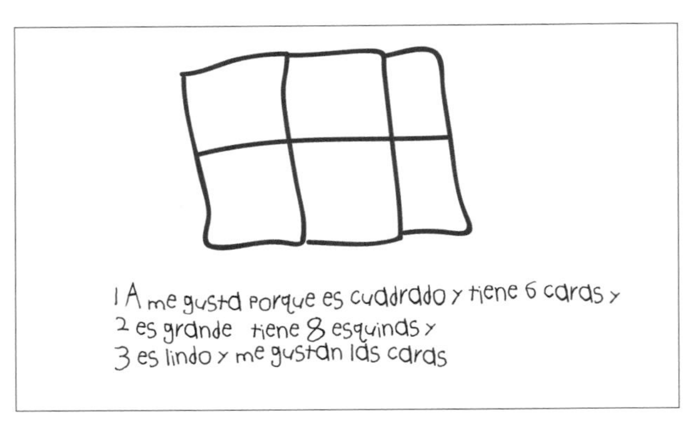

Fig. 6.13. Brenda's description of a cube

Anthony had a rectangular prism, and Rosa, another cube. Both Anthony and Rosa had correctly counted the number of faces and identified the shapes represented on their blocks. In attempting to draw their blocks, however, neither student had felt limited to the number of actual faces or shapes (see figs. 6.14 and 6.15). 375

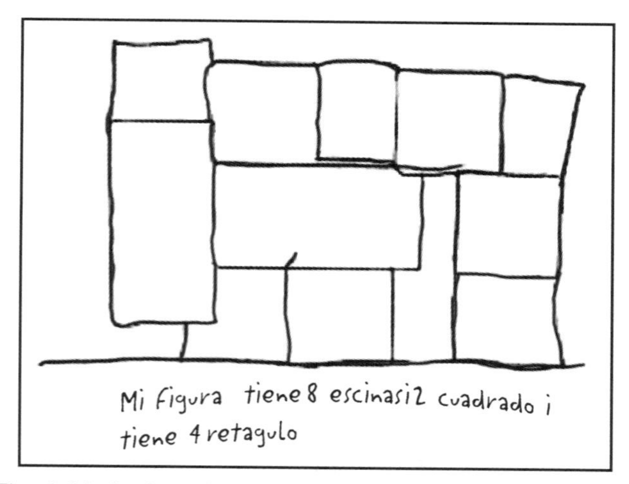

Fig. 6.14. Anthony's description of a rectangular prism

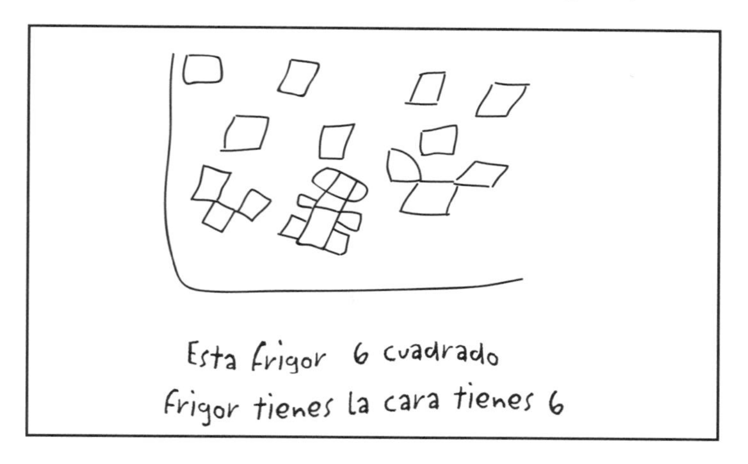

Fig. 6.15. Rosa's description of a cube

Anthony accurately described his block as having 8 corners, 2 squares, and 4 rectangles. His picture, however, shows 2 rectangles and many squares, arranged so they are touching the sides of the rectangles and seem to surround them. Similarly, Rosa's paper correctly identifies 6 squares and 6 faces for her cube, while her drawing shows more than 20 squares. In looking closely at what they drew, it occurred to me that, like Brenda, they were attempting to show the relationship of the faces to each other. Anthony's squares and rectangles all touch, and similarly, Rosa's three drawings in the bottom row seem to show how these faces are connected or attached. I wonder if some of those "extra" squares are her early efforts at trying to show the way the faces are touching. The largest configuration she has drawn, although it has too many squares to be accurate, does look something like a cube flattened out. 380 385

Alondra had been given a triangular prism like Paco's. She had correctly counted the number of triangular and rectangular faces (see fig, 6.15). Unlike the other students, she did not represent all the different types of shapes on her block, but drew just two triangles. She wrote, "It has three corners [esquinas].” Some of her classmates had been using the word *esquina* to refer to the vertices of the 3-D blocks, but Alondra seemed to be looking only at the three corners of one triangular face.

Fig. 6.15. Alondra’s description of a triangular prism

Jessica also worked with a triangular prism. She drew a single shape, yet still managed to suggest the 3-D aspect of the block. Her statements include one referring to the size of the block: “(1) My figure is large. (2) It has two triangles. (3) It has 6 corners.” I noted how dominant the triangular shape is in her picture. I also found it interesting that she showed how the block was 3-D by drawing in just one extra line. It reminded me of pictures I had seen of a piece of pie. Her picture didn’t, however, show the six points she mentioned in her description. I wondered if she noticed that.

Fig. 6.16. Jessica’s description of a triangular prism

Graciela’s and Isaiah’s drawings of their blocks appear exactly the same—both drew a rectangle. However, Graciela was given a rectangular prism, and Isaiah a triangular prism.

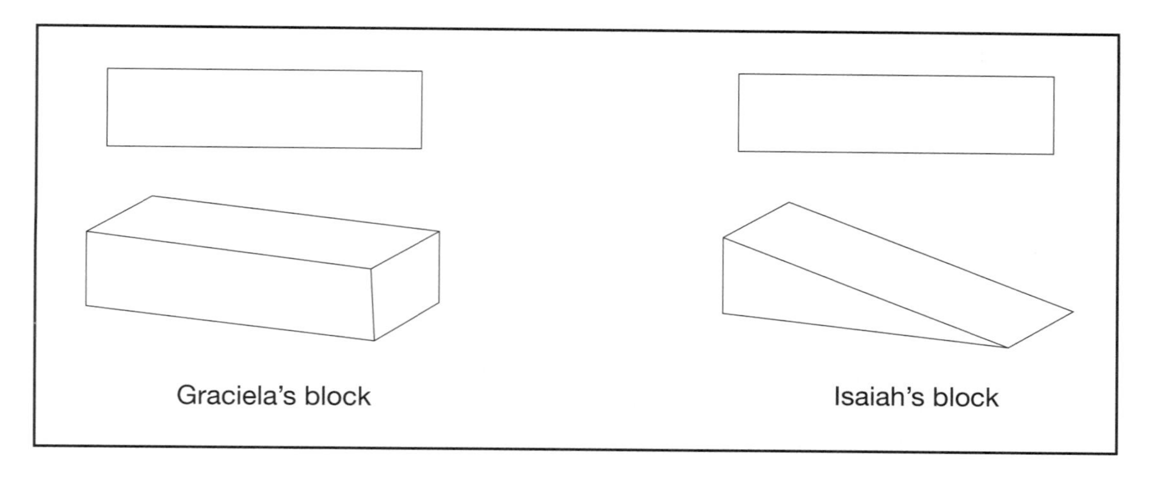

Fig. 6.17a. Graciela's block

Fig. 6.17b. Isaiah's block

I'd like to discuss these last two examples with the class. Perhaps tomorrow, I'll show both the drawings and the blocks and ask how the differences between the two blocks could have been represented. That might be a good way to continue this work. I am impressed with how seriously the students took this task and with the thoughtfulness they brought to it. Their work has given me much to think about as I ponder where to go next. 405

case 29

Boxes

Anita

GRADES 3 AND 4, FEBRUARY–APRIL

I am convinced that we need to provide students with more experiences focused on the shape of things. While we live in a 3-D world, most elementary math is 2-D. I am very interested in how children make sense of 3-D shapes and the 2-D drawings we use to represent them. How do they perceive 3-D objects? What characteristics of the solid shapes do they attend to? How do they construct a 3-D shape from a 2-D drawing? How do they represent a solid object on paper? What do they learn about the shapes by working on both kinds of tasks?

This year I decided to approach these questions with my class by undertaking a series of lessons based on designing boxes. My plan was to revisit the subject at intervals of 3 to 4 weeks. This case is the story of three such lessons.

February: Which Patterns Make a Box?

In today's geometry activity, we discussed how to make a pattern of squares that will fold up to make an open box that will hold one cube. The rules for making a pattern specify that it must be made from a single piece of paper; it can be folded only on the drawn line segments; and no sides can overlap. The class looked at a variety of patterns (see fig. 6.18) and deciding which ones will fold up to make a box (from M. Battista and D. Clements, Exploring Solids and Boxes, a grade 3 unit of *Investigations in Number, Data, and Space*, Scott Foresman 1998). I noticed that most of the children looked at the patterns and wrote yes or no after simply glancing at them.

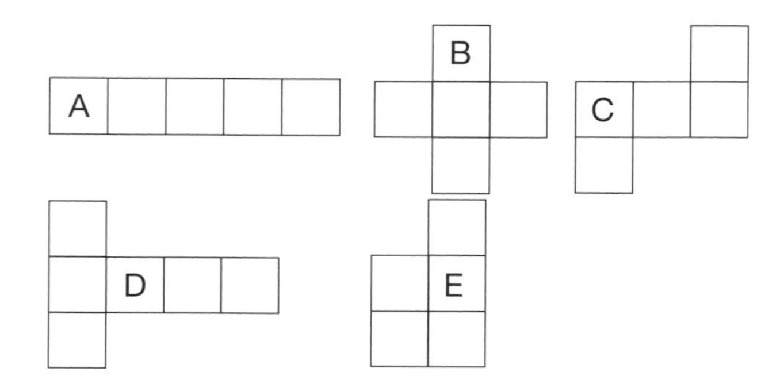

Fig. 6.18. Which of these patterns fold into an open box?

TEACHER: Tell me how you figured these out.

DAVID: I folded the shapes in my mind.

I am intrigued by this image. How do kids hold a 2-D figure in their head and then mentally manipulate it into a 3-D object? The other children seemed to like David's phrase. This idea of "folding things in your mind" is one they share. There is general agreement about which shapes would make a box, but they find it difficult to articulate why some patterns work and some do 430 not. Unable to explain their process, they instead use gestures or point in the air.

Toward the end of the discussion, Krista noted that shape A, which cannot be folded to make a box, is an array and that none of the other patterns are arrays. (We had studied arrays, rectangular arrangements of squares, during our work on multiplication.) Krista's question makes me wonder if any other students are thinking about box patterns and arrays. I decided 435 to pursue this question. I took six of the array cards we had used for our multiplication unit and displayed them in front of the class.

TEACHER: Is there an array that is also a box pattern? Let's look at some of the arrays we used in our work on multiplication to see what we notice.

440

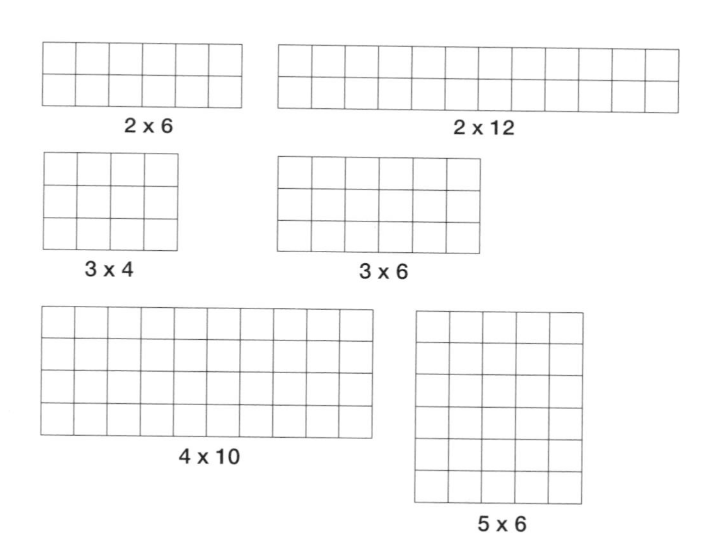

Fig. 6.19. Different arrays

They look at the arrays carefully. In their quiet concentration, they are (it seems) folding them in their minds.

PAT: [*excitedly*] I think I figured out how to make one! . . . [*after a pause, somewhat deflated*] Never mind.

Art picks up the 4-by-10 array and starts to fold it. 445

ART: Well, can you chop off the sides?

KRISTA: Then it won't be an array.

LEON: Well, then can you cut some of it? Maybe not a whole square?

I tell him to try it. He cuts and folds.

LEON: I would have to overlap some squares. 450

ART: You could make a box . . . by . . . maybe . . . not folding on the lines.

He tries this, but the lines don't seem to matter much. He still can't make a box.

PATRICK: I think I might know. Well, not the answer, but I think it's no. Because if you fold the sides up, then you can't fold the ends . . . [*Explaining his idea is hard, but he continues.*] And if you fold the ends up, you can't fold the sides up. 455

Patrick has come up with a very reasoned explanation. I wonder what the other students think about it. I ask again if they think an array can be a box pattern. The answers range from "I pretty much don't think so" to "No." Krista is thinking hard.

KRISTA: When I think of an array, the corners always get in the way.

There is general agreement that none of the arrays we are looking at could be boxes. However, it is a big step from that notion to being totally convinced that no array could ever be a box pattern. The absoluteness of that idea is hard. While the notion of proof is somewhat elusive to them, I can see they are beginning to notice that some 2-D drawings can be folded up to make a 3-D object and some cannot. 460

March: What Does a Box Look Like Unfolded?

Several weeks later, we examined the question in reverse: What happens when you take a 3-D box and unfold it? What kind of 2-D image does it make? We began with open paper boxes that we previously taped together to hold cubes. The students examined them and drew what they think each box should look like when unfolded. Then they flattened each box to compare it with their drawing. 465

Our routine at a local food pantry offered yet another opportunity to explore the situation: one job involved taking apart boxes to prepare them for recycling. The first box we examined was a shallow, open box that held cans of tuna. I saw that its design was similar to the paper boxes we made in class from simple patterns. 470

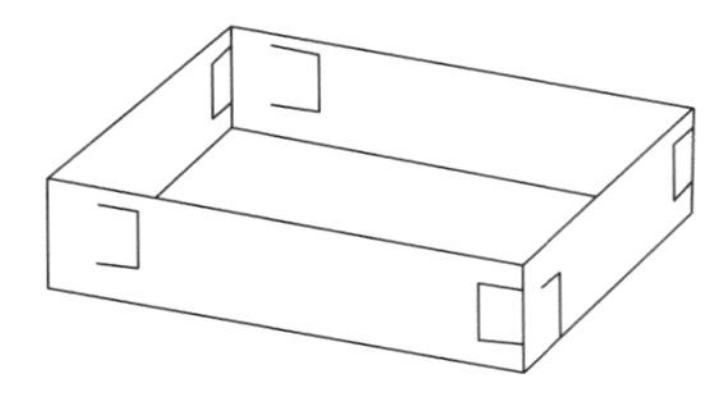

TEACHER: Before we flatten this box, I'd like you to look at it and think about what it will look like unfolded. 475

ART: Is this supposed to be for you to figure out how we unfold things in our mind? When I look at it, I try to think about each part. These two would come apart. [*As he speaks, he points to the left- and right-hand sides of the box.*]

DAVID: This would come down to here. 480

I thought Art and David were saying the sides of the box would flip down and out to the sides.

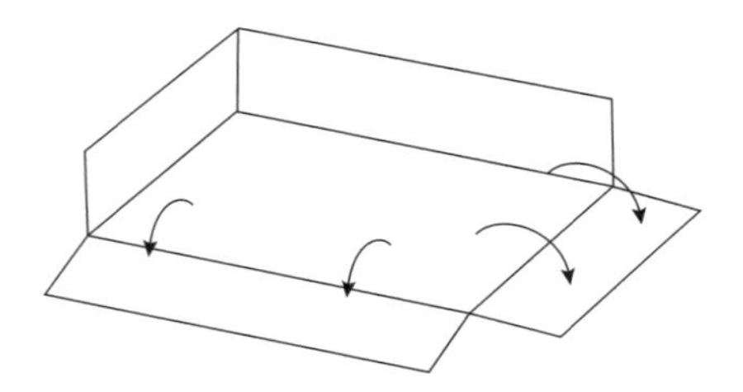

TEACHER: What else do you see?

SUMANA: The sides aren't that high.

PATRICK: I noticed that at the bottom of each side there is a small hole. 485

KRISTA: There's tape that holds the sides.

When we took it apart, the shape was surprisingly familiar.

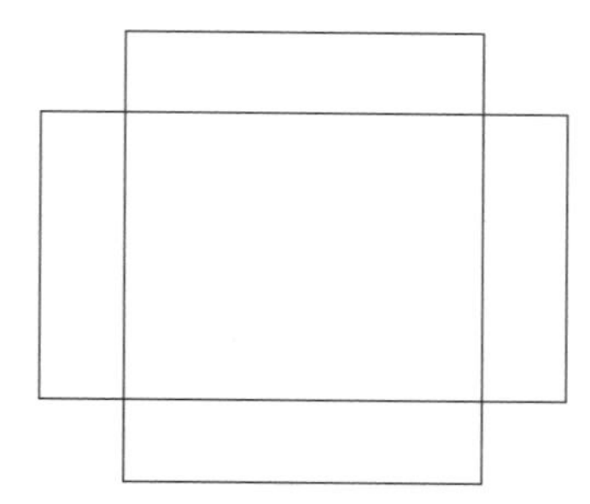

KRISTA: It's like the box pattern we made in class!

ART: This is the same thing, except the middle is wider and the sides are shorter. 490

The next box was more complicated in design, not only in the way it unfolds but also because there were extra flaps that are used to hold the box together.

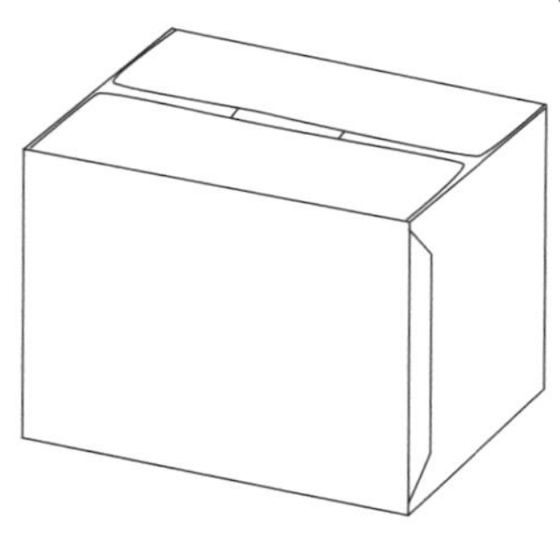

ART:	Are we going to study it first?	
SUMANA:	Are you going to flatten it? Like, totally make it flat?	495
TEACHER:	Yes, totally flat. First, what do you see?	
PATRICK:	It has flaps on top.	
KRISTA:	It's bigger than the other box. It's a closed box.	
DAVID:	It has little holes like slits.	
SUMANA:	There's a hole in the bottom.	500
KRISTA:	It would be like having sides on it and corners.	
PATRICK:	It has a top and a bottom.	
ART:	It's raised in its width. It's high and wide.	

They seemed ready and anxious to draw this box flat, and so they began. At first they (505) didn't look at each other 's drawings because they were too busy sorting out their own ideas. I wondered if I had given them a challenge that would confuse them. This design was significantly different from the open box patterns. After a while, the children began to look at each other's drawings (Some are shown in fig. 6.20). They noticed the differences and compared their interpretations. (510)

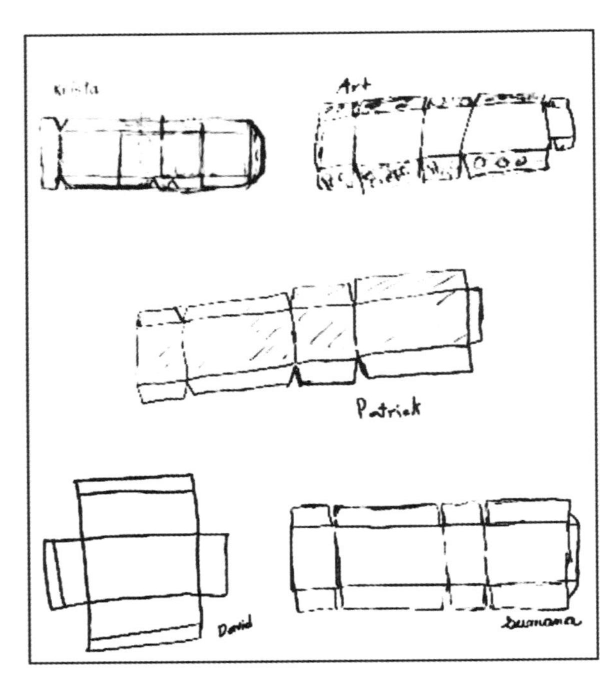

Fig. 6.20. Many of the children had very similar ideas about this box.

When we unfolded the box, this is what we got:

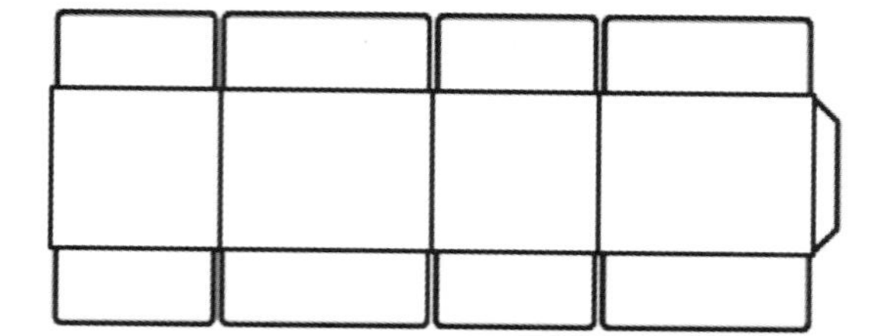

As the class looked carefully at the flattened box, there were some exclamations of surprise. They were almost talking to themselves, checking and looking at the different big and small parts. 515

DAVID: I notice that I did it a different way.

David's drawing was the most like the open box we had done earlier. Perhaps he believed that all boxes would fit his pattern. On the other hand, he was interested to see that the others approached the task differently. I noted that even those who drew similar patterns had different ways to show the flaps. In addition, Patrick, Krista, and Art used shading to represent the different sections. 520

PATRICK: I was right. I made the tab on that side.

They really liked to fold and unfold the boxes and compare the results with their drawings. It was like solving the puzzle of the box design. I think I'll have them design their own boxes when we revisit this work in a few weeks. 525

April: Design-Your-Own Box Patterns

I began this lesson by telling the class they would have a chance to design boxes of their own. I had thought about asking them to make a box with more than six sides or limiting them in some way. Instead, I decided to keep it very open. They began by thinking and drawing. Art, Patrick, and Krista each did something different. 530

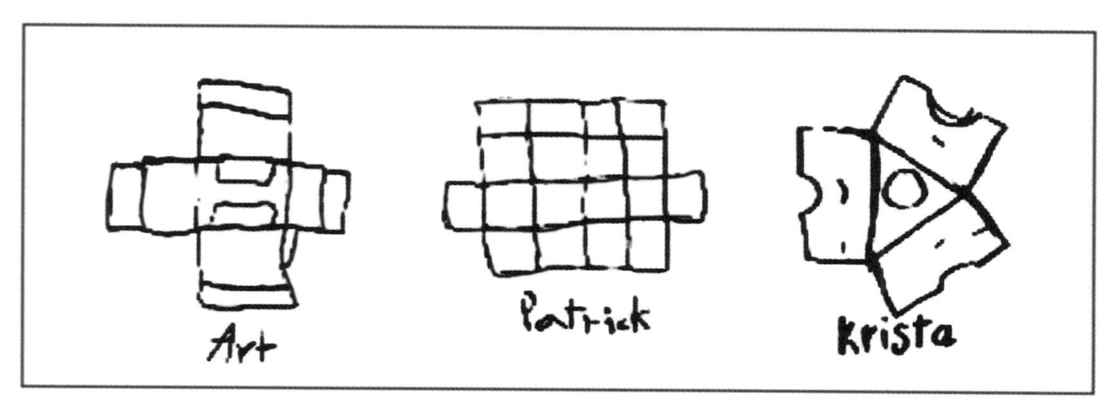

TEACHER: What do you think about when you are making these designs?

ART: The way I want to do it. I want to make a box when you unfold it, there will be four little boxes within one box. The top would go like ahead across the paper. 535

PATRICK: I want to make a box that would be sort of tall and thin. I did those lines to see where the folds went. It can have cubes inside.

KRISTA: I had the idea of a triangle in my head. I got a soccer ball for my birthday and it came in a triangle box.

PATRICK: Now I'm going to make another one. I want to make a round box. I got the idea from the pencil sharpener. 540

They continued to work on their designs. I checked in with Patrick. I noted that he was fitting a strip of paper to serve as the side of the box around a circle. The circle was the bottom of his box. I asked him how he got the strip of paper to fit around the circle so perfectly.

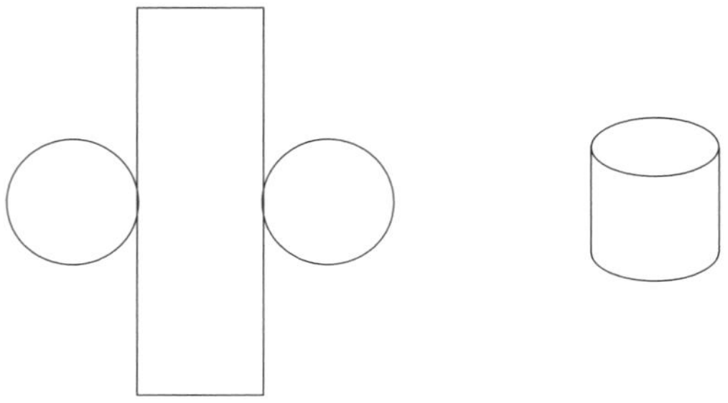

545

PATRICK: Oh, I cut a strip, and then I made it into a circle. Then I traced the circle and cut it out. Then I needed to make one more. I thought it wouldn't work, but then I thought it would work. It works.

KRISTA: I started with a tiny triangle box. [*Indeed it is very tiny—less than an inch high.*] The first draft was much harder. Now it's easy. Before I made a triangle; this time I made a diamond. 550

Krista was busy folding her box, but I asked her to show me what it looks like unfolded. I wanted to understand what she means by "diamond." When she opened up the box for me, I saw that the two triangular faces and the edge that they shared did make a shape someone might call a diamond. 555

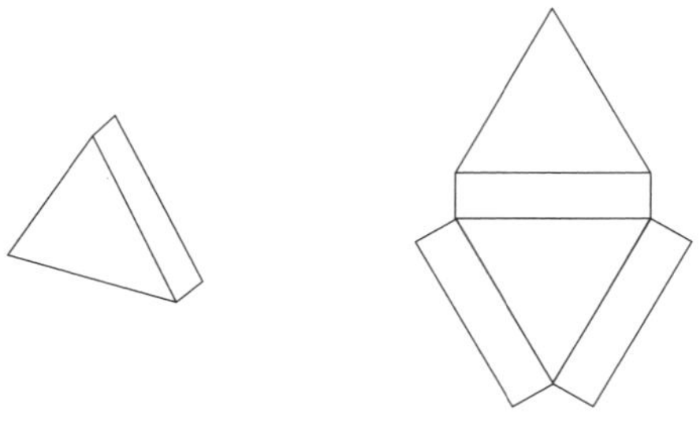

KRISTA: The edges are a little small. Next time I would make them bigger.

All three of these box activities have helped my students look carefully at the relationship between 3-D objects and 2-D representation of those objects. Their ability to move back and forth between the different dimensions still presents a challenge, but their flexibility is significant and I see lots of progress. 560

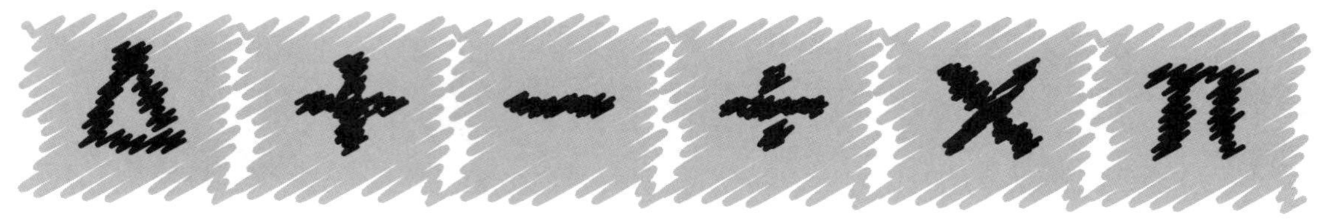

CHAPTER

7

Reasoning in geometric contexts

Children call upon their geometric knowledge in a variety of settings and problem-solving contexts. While using geometrical ideas to solve problems, children extend their knowledge of geometry as well as their ideas about what it means to reason mathematically. The cases in chapter 7 provide examples of children using geometry as they solve both mathematical and nonmathematical problems.

In the first two cases, two fifth-grade classes explore how the concept of *middle* might apply to noncircular shapes. What mathematical ideas are they developing in their work? In the third case of the chapter, we encounter a seventh-grade student, Masha, whose work with the area model for fractions highlights the relationships between figures that are congruent and figures that are not congruent but have the same area. And in the last case, we examine the thinking of Janine's fourth graders as they create mathematical arguments to solve a perimeter problem. What math ideas do they call upon? What do they understand and not understand about mathematical argument?

case 30

Finding the center: X marks the spot

Jordan

GRADE 5, FEBRUARY

Our school librarian gives out awards for classes that return 100 percent of their library books on time each week. This year, my class is very interested in getting that little piece of paper. Looking for a teachable moment and coinciding with our study of geography, I suggested that we identify each wall in our classroom as the north, south, east, or west wall, and then aim to get a library award for each wall. The race was on.

Before long, we had our four awards. The class had a good system for getting their books in on time, and we all knew it was just a matter of time before we had a fifth award. Where in the room would we place that award? Suggestions flew: (1) one of the corners, (2) a place that would represent a direction like southeast, (3) the ceiling, (4) the floor, or (5) dead center! We took a class vote and it was official: The fifth award would go on the floor in the exact center of the room. And where was that? The excitement to find out was high. First we predicted, simply by eyeballing, where the center of the room was. Then I pushed for a more accurate reckoning.

TEACHER: How could we figure out where the exact center of the room is? How can we prove that it is the center of the room?

The class spent some time discussing this within small groups and then gathered to share their ideas.

DONTE: A student stands in the middle of each wall. They all start to walk forward toward the center. Where they bump into each other is the center.

Some of Donte's classmates asked him about measuring the space with footsteps. They remembered a previous activity in which it was clear that twenty-five different-size feet altered the measurement. In addition, they felt clarification was needed for how to find the middle of each wall.

LORI: This is like Donte's idea, but I think you just need two walls. Using the floor tiles, just find the middle of this wall and find the middle of that [adjacent] wall. Have two walkers walk toward the center, counting one tile for one step. They should reach the center in the same amount of tiles.

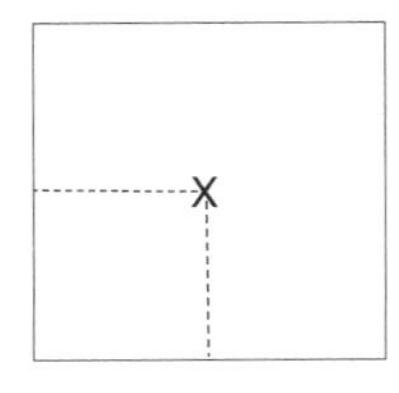

MAC: I think we should take string across the room both ways.

TEACHER: Mac, can you explain this more so we can get a picture in our mind of what this looks like? 30

MAC: [*coming to the board to sketch his idea*] Take the two strings like this.

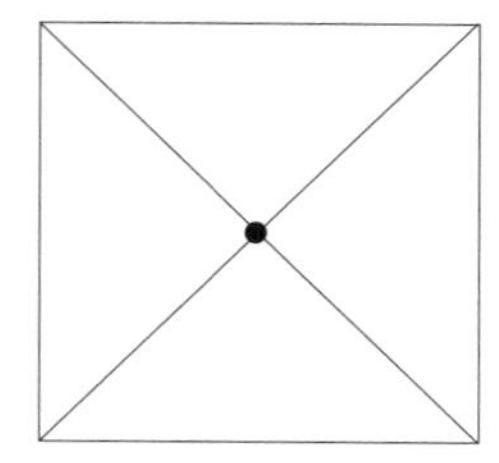

TEACHER: Oh, so your suggestion is to stretch the string from corner to corner in a diagonal way?

MAC: Yes. Maybe you could tape it to the wall or somebody could hold it. 35

TEACHER: And the center?

MAC: Right about here [*points*], where the string crosses the other string.

For homework, I invited them to think about this further and come in with a solid proposal for tomorrow's lesson. I was wondering if anyone noticed the difference between Lori's and Mac's methods. Lori's depended on knowing the room was square, while Mac's would work for any 40 kind of rectangular shape. In Lori's method, they would count the tiles and the count would be equal only in a square. If she had been talking about finding the point where the two paths cross, her method would also work for rectangles.

The next day we broke into groups to investigate. Kenny's group asked to use the spare room across the hall because there was less furniture in it. I agreed, but asked what method they 45 planned to use.

KENNY: We're going to count ceiling tiles.

TEACHER: Can you tell me more about that?

KENNY: Well, the ceiling would fit on the floor if the walls weren't holding it up. So you can get the same results by using the ceiling. 50

TEACHER: So the ceiling and floor fit each other?

KENNY: Yeah, and it's easier to count the ceiling tiles, because they're bigger—too many tiles on the floor.

TEACHER: Anyone working on the troubleshooting?

GRACE: We have to make sure that the spare room and our classroom are the same 55 size or the center will be off.

Grace also commented that some of the tiles weren't full-size.

TEACHER: How will your group work with that?

GRACE: I think maybe the two tiles on the ends might be half, so it might even out.

As the groups bustled around the room, I noticed that no group's proposal included rulers, tape measures, or any standard measurement tools. They were looking at the space in the classroom in a very uninhibited way. No numbers, no formulas! What they seemed to be noticing was shapes and space. 60

As the groups finished, they placed sticky notes on the floor to mark *their* center. One group insisted on putting their sticky note on the ceiling because . . . well, it was the same! Otherwise, all the little yellow squares ended up on the same tile, within a few centimeters from one another. As Kenny exclaimed, "X marks the spot!" 65

When the students returned to their seats, I gave them time to write personal reflections, defining center and explaining how they could prove that their spot was the center.

The next day we worked with graph-paper patterns of our floor and continued to explore how to find a center point. I was amazed at the variety of math ideas that grew from this work. For instance, Kayla said that the opposite sides of the room had to be equal in order to find the center, but if you "slant" the shape of the room, the center is not the center anymore. I am thinking this is the same Kayla who announced to me in September that she couldn't do math. Yet in this class, she is the one to mention that the method they are using depends on two factors—the opposite sides need to be the same length, and the angles must be right angles (or not "slanted" as she put it). Other student comments included these: 70 75

"You have to know about the space inside the shape."

"You can find the center of anything using straight lines."

"We learned that you might have to try a lot of different ways to find the center, but just keep trying and you will get it." 80

A few days later, the librarian came to our room to present another award for having all our books returned this week. Here we go again!

case 31

Does a triangle have a middle?

Rita Lucia
GRADE 5, MAY

I was working with a group of three students who were discussing geometric shapes. I asked them to share some examples of these shapes from everyday life. Efren offered the sundial as an example of a circle, then thought a minute and added, "Actually, a sundial is a triangle standing on its edge in the middle of a circle." The other two boys in his group added that clocks and pizza pans are also circles. 85

When I suggested large circular restaurant trays as another example of circles, this began a conversation about "balancing the trays with your hand underneath and sort of in the middle." Leo noted that clocks have hands connected in the middle. Orlando said pizza pans don't have a middle, but Leo disagreed: "When you get a pizza and it's cut into slices, where all the cuts cross is the middle." Throughout the discussion, no one questioned what the word *middle* really means, and I decided to see what they might think about *middle* if the shape isn't a circle. 90

We began by cutting out a variety of geometric shapes from cardboard. First, a circle. Where's the middle? Immediately, they put their fingers in the center. I presumed they were using the eyeball method. 95

TEACHER: How can you be sure?

LEO: If you measure from where your finger is to the edge of the circle in lots of different places around the circle, it should be the same distance. If they aren't all the same, then you move your finger over a little and measure again. 100

TEACHER: Would you call that "guess and check"?

LEO: Yeah, you keep guessing where the exact middle of the circle is and keep measuring.

TEACHER: Can anyone suggest a way to confirm what Leo thinks? 105

EFREN: If you measure all these lines from the edge to that dot in the middle, and they are the same, then you know that dot is in the middle. If they aren't the same, then you have to put the dot someplace else.

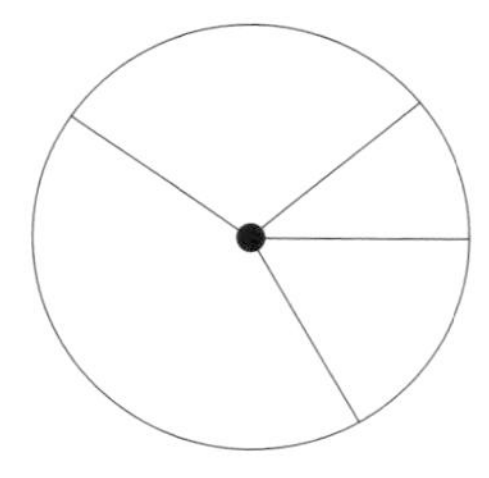

I noticed that both boys were thinking about the radius of the circle and using that as the basis of their strategies. I expected that someone might draw lines across the circle, as if slicing up a pizza. No one did. Also, no one suggested measuring across the circle and checking to see if the dot was at the halfway point. I placed my finger on the point they had indicated as the middle to see if it would balance. The cardboard circle balanced on my finger.

TEACHER: Can we use this as an additional proof of the location of the middle?

They agreed, and all of them wanted to have a try at balancing it. I saw their certainty about the middle of the circle.

TEACHER: What about the square? Is there a middle to a square? Do you use the same procedure to find the middle of a square?

Efren seemed sure he had a way to find the middle. He made the following diagram to prove his thinking and measured the segments with a ruler.

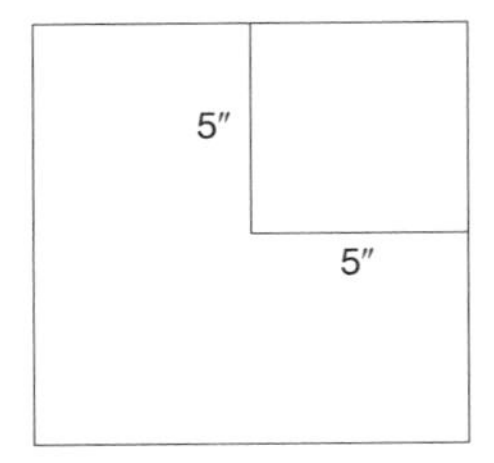

TEACHER: Convince us that the dot is in the middle of the square.

EFREN: This is the middle. Both lines measure five. They are the same, so that's the middle. The lines also make a right angle.

TEACHER: Do you mean whenever you measure from the side of the square to the dot and it measures the same number of inches, it proves the dot is in the middle of the square?

EFREN: No. Here is a diagonal. See, it is 7 inches, but that is OK. It is a diagonal and diagonals are always longer. If two boys walk to the same spot, and one walks in a straight line and the other walks in a diagonal line, the diagonal will be farther. This one measures 5 inches and that one is 7 inches.

Because Efren is measuring this diagonal, he can get only an estimate of its length. The actual length is so close to 7 that I decide to let him continue his line of thinking without commenting.

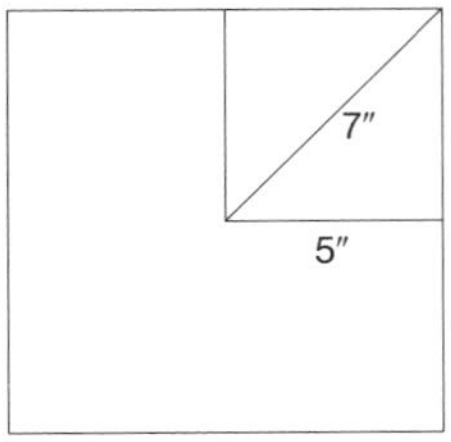

LEO: It's OK. If you draw the other side of the diagonal, and it measures 7 inches, then it's OK. You have the middle.

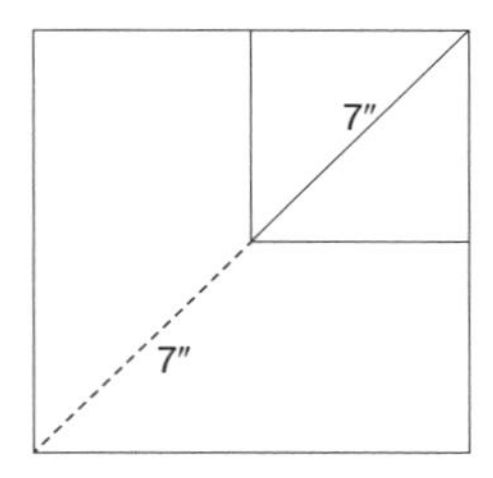

EFREN: So the straight lines are the same length and the diagonals are the same. Now you know that you have the middle.

LEO: Draw in all the other parts of the lines and measure them all. If they are 7 and 5 inches, then we know we are right, and that is the middle. 140

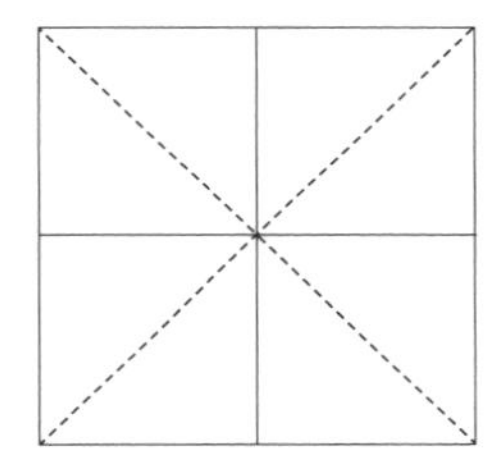

They were relying on their understanding of linear measurement to find the middle of the square. As additional support, Efren used the fact that the two lines met in a right angle. There is something neat and tidy about a right angle. "How can these two lines, drawn from the midpoints of two sides, which meet in a right angle, not point the way to the middle?" thought 145 Efren. Everyone wanted to apply the "balance" test to the square. Once they balanced the square on one finger, they accepted this as further proof that they knew how to find the middle of a square.

I wanted to continue to push their thinking, so I picked up a cardboard piece that had the shape of a scalene triangle. 150

TEACHER: Is there a middle to a triangle?

Orlando's thoughts spill out, one after the other, most of them incomplete. It is fun listening to his ideas as they are being formed.

ORLANDO: No, no, there is no middle. You can't measure it. It doesn't have four sides or no sides, it has three sides. . . . I guess there could be a middle, but there is 155 no way to prove it. There is no way to measure it. I can't see where we can measure.

EFREN: I think there is a middle. [*He uses a ruler to find the midpoints of two of the sides.*] That's the exact half, in other words, where the middle usually is.

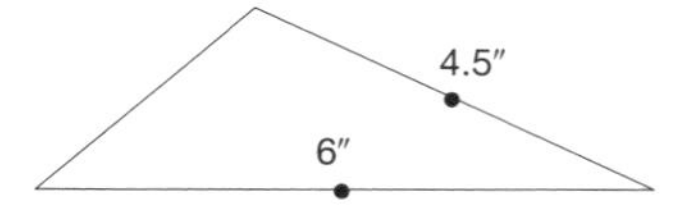

TEACHER: Can you clarify that for me? What do you mean? 160

EFREN: The middle is usually half of things. So if I measure a side and it is 12 inches long, then the middle of the side will be at 6 inches. Middles are half of things. [*He draws a line segment from each midpoint until the segments meet.*] They aren't the same length, but that is OK. The dot is in the middle. This is the middle. That is where the lines meet. That's the middle. I bring them 165 together here.

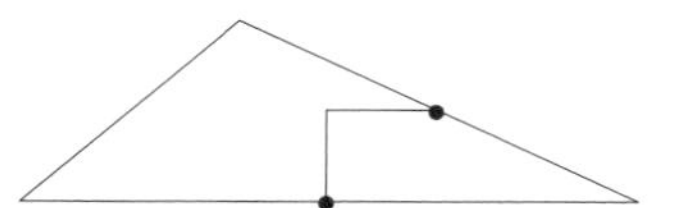

Efren relied on the strategy that worked for the square. It looked so right: two midpoints and two lines that met in a right angle. He was sure that this was correct. Efren quickly picked up the triangle and tried to balance it at his "middle" dot. It fell.

EFREN: This is not the middle. 170

Everyone had to try balancing the triangle on his or her finger. I wondered about their need to do the balancing each time. Was this just fun, or does everybody need to experience the proof for themselves? Everyone agrees that this was not the middle.

Efren still didn't want to believe it. He said that there was something wrong with the way we made the cardboard triangle. I think he was so convinced his method worked that he wanted 175 to find some other explanation. He looked carefully at the edges of the cardboard to see if the design was faulty. He wondered out loud, "Maybe it has something to do with the way you cut the cardboard."

I was thinking that it doesn't hurt to explore all avenues before abandoning a strategy. Meanwhile, Orlando was still trying to balance the triangle by trial and error. He succeeded and 180 said he's found the middle.

TEACHER: Where is the middle?

LEO: Under his finger. That is why it is balancing. Let's look under his finger.

I thought Leo wanted to look under Orlando's finger to see if Efren's middle dot was there. Leo knew that if Orlando balanced the triangle, he must have found the middle. They saw that 185 Efren's strategy didn't work, but it did seem reasonable to them, and they didn't know what else to do. When we turned the triangle over to look under Orlando's finger, we saw that Efren's dot was nowhere near Orlando's finger.

They decided to revise Efren's method by measuring to find all three midpoints. Efren connected two midpoints with a line and then drew a line up from the bottom point until the two lines meet. He announced that this point was the middle. 190

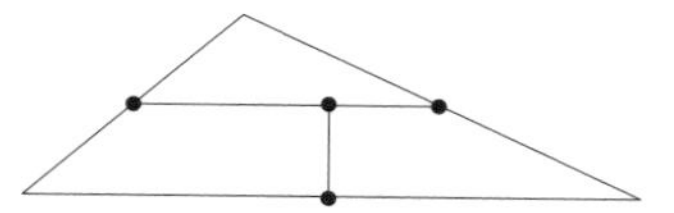

When Leo measured and found out that the three segments were not the same length, he looked puzzled. Efren attempted to balance the triangle on his finger, but it still didn't balance. "I'm stuck," he admitted.

Orlando picked up the pencil, erased Efren's lines, and connected the dots in a triangular shape. 195

ORLANDO: The middle is in the middle of this triangle. But that is the same problem, trying to find the middle of a triangle.

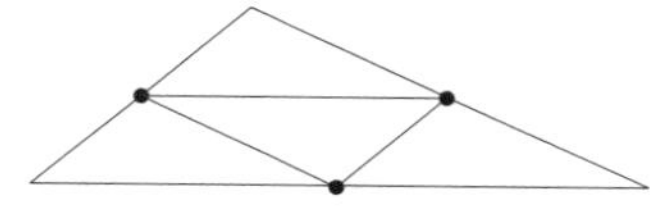

TEACHER: This is a really good problem, but it is hard. Do you think this means there might not be a middle to a triangle? 200

Everyone cried out in disbelief at my statement. I could tell by this reaction that they were really interested in solving this problem and wanted to continue.

LEO: I think there is a middle. We just can't find it right now.

TEACHER: You have found the three midpoints of the sides. Then you saw that connecting them together did not give you the balance point. What else might you do with the points? 205

Efren almost exploded with a new thought. He erased all the work on the cardboard except the midpoints. He drew a line from the vertex of one angle to the opposite midpoint dot. He did this for two angles only. Where the lines intersected he made a dot. 210

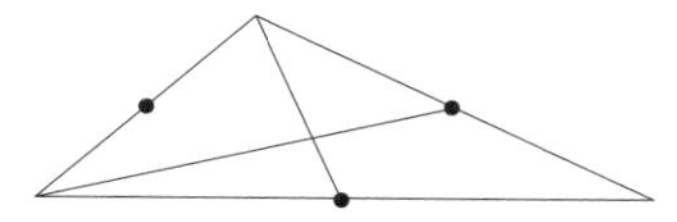

EFREN: This is the middle.

Everyone clamored to try balancing the triangle at this dot, but Efren prevailed. It balanced!

TEACHER: Why did you draw lines from two vertices? In your first attempt at measuring the triangle, you measured two sides. In this strategy you are making just two lines. 215

EFREN: I did it to the square and it worked. I measured two sides and marked half of each one, and then drew the lines in.

TEACHER: Let's think of what makes a shape a triangle. What are the characteristics of all triangles?

The boys concurred that all triangles have three straight sides and three angles. This was 220 enough of a suggestion for Efren. He erased everything on the cardboard and remeasured all three sides, making a dot at the midpoint of each. Then he drew a line from the vertex of each angle to the midpoint on the opposite side. Where the lines intersected, he drew another dot and announced, "This is THE MIDDLE."

When Leo measured from the dot to the vertices, he found they weren't equal. The boys 225 were somewhat puzzled by this, but still convinced by the balancing method that this was the middle of the triangle. They tested this method on the other triangles and found that it worked for all of them.

As I reflect on this investigation, I note the way the boys took the procedures that worked with circles and squares and kept modifying them to make them fit the triangular shape. Finding 230 the halfway points of the sides of the shapes was an important aspect of locating the middle for these boys. The balancing test holds so much meaning for them that they simply gave up the other test they had been using—that the line segments from the middle to the vertices had to be the same length. Without realizing it explicitly, they had changed their very definition of middle as they did this work. 235

Is there such a thing as a "middle" of a triangle? I remember studying several kinds of lines related to triangles, and some of these three students' approaches are reminiscent of that work. For example, one of their rejected strategies was based on bisecting the angles, and another started with drawing perpendicular lines at the midpoints. I recall that when I was in school, we examined the ways lines such as these intersected, and we considered how the intersection 240 point would change when the shape of the triangle changed. While that work isn't appropriate for my fifth graders, it was interesting to see where their curiosity about "middles" took them. They certainly had a chance to work with a lot of geometry as they looked for the middle of a triangle!

Would I want to suggest that there could be more than one middle or that points might be 245 different kinds of centers? Since their sense of middle is now based on balancing, what kind of sense would these other points have for them?

Early in the discussion, there was a lot of talk about distances needing to be equal. I wonder how would they react to the point in a triangle that is formed by line segments that are equally distant from the vertices? Would they call that the middle, even if it isn't the balancing point? 250 That seems like a good investigation to do next.

case 32

Are the areas equal?

Sandra
GRADE 7, FEBRUARY

This year my students and I have used area as a model for understanding fractions. We made fraction bars that we then used to explore comparing and operating with fractions. (Fraction bars are simply long rectangles marked into congruent sections to represent halves, thirds, quarters, and so forth.) In a wonderful coincidence, I was recently flipping through an old issue of *Mathematics Teaching in the Middle School* and saw an article about assessing student understanding of fraction ideas using areas within a square. ("Using Cases to Integrate Assessment and Instruction" by S. K. Wilcox and P. E. Lanier, January 1999, *Mathematics Teaching in the Middle School* 4 (4), pp. 232–241. Activity adapted from the Balanced Assessment for the Curriculum Project.) The assignment asked students to identify fraction names for all the labeled sections in the following square: 255 260

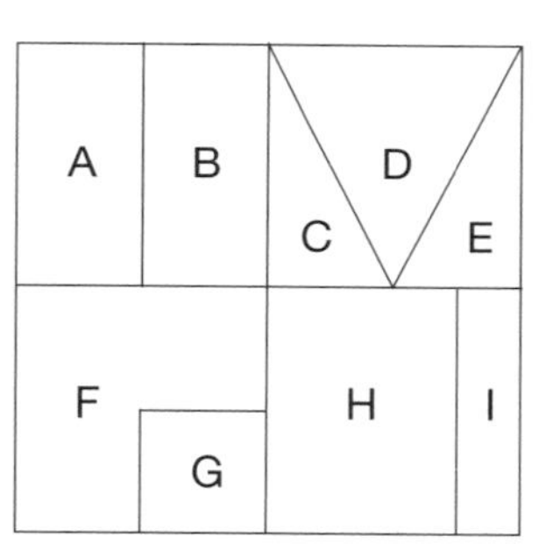

Some of the pieces had the same area but were clearly not congruent. I wondered what my students would do with this problem—in particular, with the different-shaped pieces of equal area. Would they see that pieces with the same fraction name have equal area, even if they are shaped differently? 265

Students worked in pairs to name the fraction for each piece in the given square. Many students first recognized that the large square was already divided into quarters; they then subdivided each quarter into quarters to form sixteen smaller squares. With this approach, piece G was an obvious choice for the starting point, and everyone agreed with the name—$1/16$. Other students started with the square divided into quarters, but then subdivided the quarters into triangles (like shapes C and E) or rectangles (like shape I), each of which was one-quarter of the quarter. 270

During our class discussion, I decided to push for explanations about pieces that seemed to have the same fraction name but clearly did not look alike. Students were quick to volunteer explanations. 275

JOEY: G and I are the same because if you split I in half and then place the two pieces side by side, you get G.

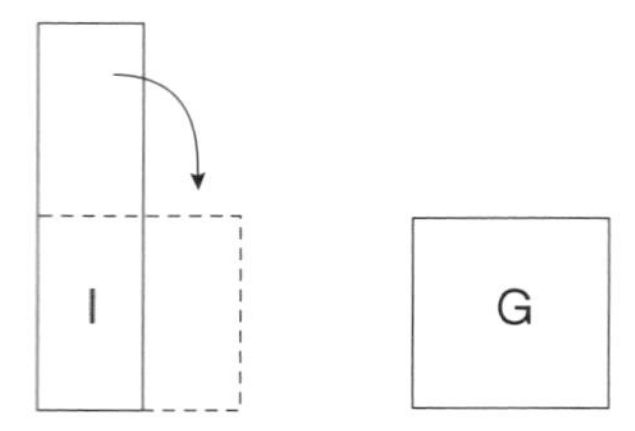

HAYLEY: I know that I and E are the same because of what we did on the geoboards. Move the triangle at the bottom of E to the top, and you get I.

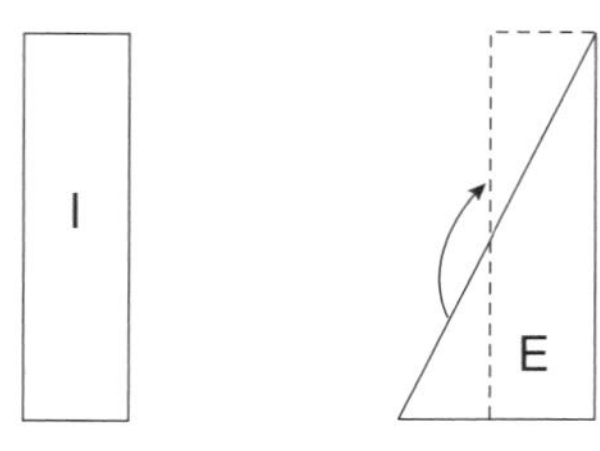

Students gave fraction names to the other pieces by breaking apart and reassembling them to make shapes congruent to the smallest starting piece. Some students used G as their unit; others used I or E. I was surprised that their explanations for naming fractions for the given design had been so easy. My students were moving pieces around to create equal areas when, in fact, the original piece was clearly not the same size and shape. They seemed to have no trouble making congruent shapes that could then be used to name the fractions. 280 285

My next step was to extend the activity by having them create a new design to make a similar but different "naming fractions" problem. I expected that this would require a different kind of thinking. Would the students remember that pieces of equal area were needed to name fractions? Would they use different shapes and convince themselves and others that the areas were the same for each shape? 290

The next day Masha came to me for individual help. She had created a design but was unsure of her fraction names.

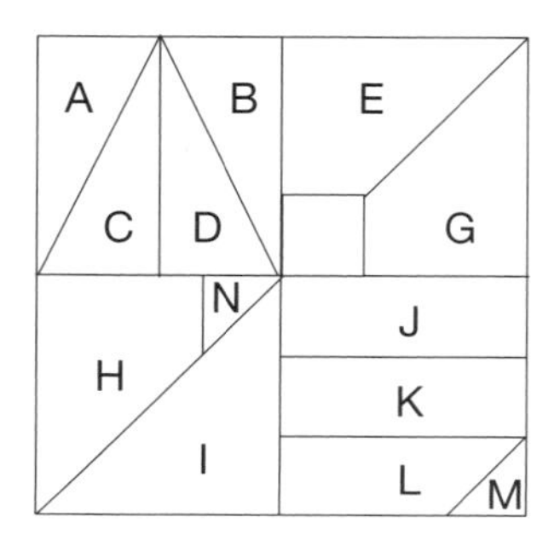

TEACHER: This is a very interesting design. How did you decide on these shapes?

MASHA: I liked the look of the triangles for A, B, C, and D, so I tried to make some other triangles of different sizes. H and I are pretty easy, and they're big. I wanted something else interesting, so I put M and N in the corners. F just happened to cover up another corner. I'm not sure how to name the fractions. 295

Masha is a hardworking, tenacious, but quiet student who likes reassurance that she's on the right track. I'm trying to make her more confident of her abilities and less dependent on what someone else says is the right way. She's a good thinker but needs time to develop her ideas. 300

TEACHER: What do you know about naming fractions?

MASHA: All the pieces have to be the same size when you break up a shape, but you can combine some small pieces to fit the large pieces and still name the fraction.

TEACHER: What does that mean? 305

MASHA: Well, if I only had this picture . . . [*drawing a square*], I would split the square into four pieces, and name these two [little pieces] each $1/4$. Then this big piece would be $2/4$ or $1/2$.

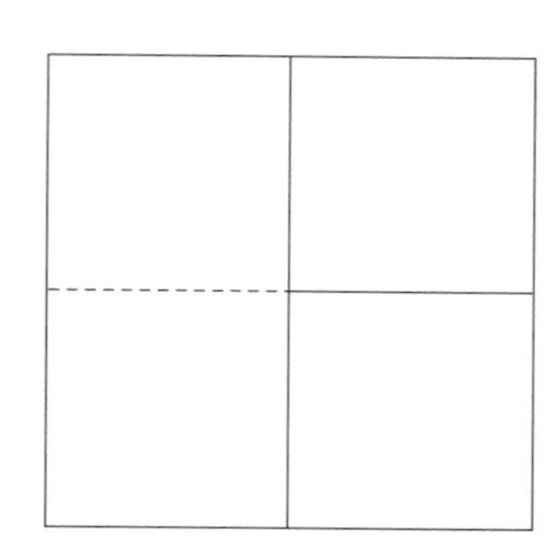

TEACHER: So, how could you use that idea to help with your big design?

MASHA: Well. I can see that M and N are the same size, and they're the smallest pieces, so I guess I should use them to try to break down the square. 310

I waited while Masha sketched a few hesitant lines in piece L.

MASHA: Hey! I can fit five triangles in L!

There was another pause as Masha was struggling with something she was not verbalizing.

TEACHER: Think out loud, Masha, so I know where you're going. 315

MASHA: I was just thinking that L and M together would be six triangles, so J, K, L, M would be . . . 18 triangles. I can't do the next part in my head, but I think I should multiply 18 times 4 to find how many triangles would be in the whole thing. I started my design by dividing the square into four equal parts, so that's why the times four. 320

I left Masha alone to continue her task. Later, when I checked on her progress, she showed me her answers to H and I as well as J, K, L, and M.

TEACHER: How did you get—8/72 for H?

MASHA: Well, you remember how I said that there were 18 triangles in this square? [*She outlines the quarter containing J, K, L, and M.*] I multiplied 18 times 4 and got 72. Then I remembered that this square [*she outlines H, I, N*] was the same size, so it has 18 triangles. Since it takes up half of the square, it must be 9 triangles. H and N together make 9 triangles, so N is 1 and H is 8. So piece H is—8/72. 325 330

There was a lot going on in Masha's thinking. She clearly understood that dividing the large square into four congruent squares created four equal areas. Later, she was able to switch to congruent triangles to name pieces J, K, L, and M, but it was quite a leap for her to identify that the original square would be 4 times the 18 triangles in J, K, L, and M. She made an even greater leap by naming piece I as 9 triangles without tracing in the actual triangles. She seemed to understand that the congruent triangles were embedded in pieces H, I, and N, even though they clearly did not match J, K, L, or M. 335

case 33

Mathematical arguments: Proving your point

Janine
GRADE 4, MAY

My fourth-grade class has done what I thought was a lot of work on area this year—area as a multiplication model, area in fractions, and so forth. I decided to challenge them a bit by having them find the area and perimeter of an irregular polygon. To make it even more interesting, I left out some of the dimensions. I suggested that they work in groups on the problem. 340

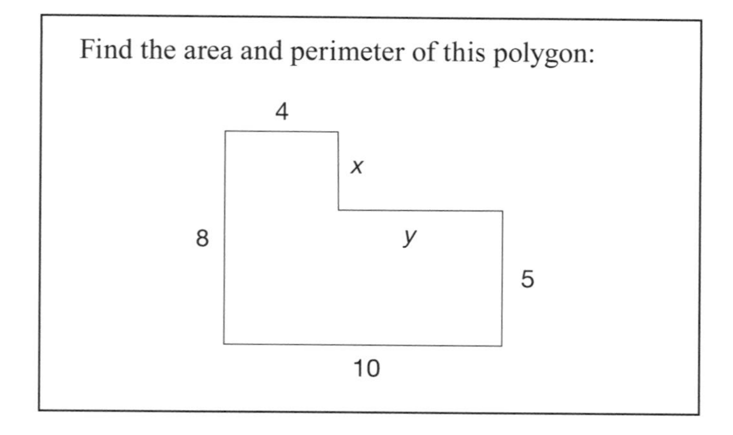

I assumed that they would either draw out the problem using the squares on grid paper or get color tiles to build a model of the shape. To my surprise, no one did either of these things. Were they seeing this as an area model even though the square units weren't drawn? Were they making any connections with what we had previously done? They took a totally different approach from what I had anticipated, an approach that caused quite a discussion. 345

All the groups set right to work. This was a challenge to them. There were missing pieces! Two segments were not labeled with their lengths. Questions arose: how can we find the perimeter if you didn't tell us how long all the sides are?

I gave them about twenty minutes to work in their groups, and then called them together to share what they had done. Anna began the discussion. 350

ANNA: We figured out what the "x line segment" was equal to. We pushed in the 5 from the other side of the rectangle (rectangle B) to the bottom of the x segment." We knew that was a 5, so we asked ourselves, "What plus 5 equals 8?" We think the x segment is equal to 3. 355

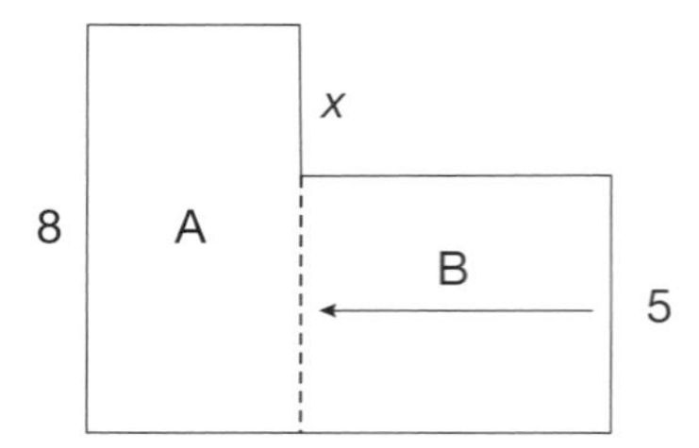

Several in the class agreed with her answer, but many looked confused! What was this "pushed in" thing she was talking about?

NOEMI: I get it. If you take the 5 line segment and push it over, it makes the other side of a rectangle. If one side is 5 and it's a rectangle, then the other side will be 5, too. 360

JUANA: Right, because they are the same size up and down. So if this one is 5 [*pointing to the left side of rectangle B*], then the top of it is a 3. This outside line segment is an 8, and if you cut off the smaller rectangle, then the other side will also be an 8, and 5 + 3 = 8! [*pause*] I think line segment x could also be a 4 because 4 + 4 = 8. If you divide up the other side, it could be 4 and 4. 365

I was very confused now. How could she make such a good case for the 5 + 3 argument and then turn around and say it could also be 4 + 4?

NOEMI: [*adamant*] Does 4 + 5 = 8?

JUANA: I'm not talking about the 5. I'm talking about the 4 + 4. 370

NOEMI: But this side [*pointing to the left side of rectangle B*] is a 5. Do you agree with that?

JUANA: I'm not sure.

ANNA: If you draw a line down. Keep drawing that line that says x until it gets to the bottom; then you can see that you have a rectangle, right? We know that the right side of the rectangle is equal to 5. If we "push" that line segment over to the left side of the rectangle, can you see that they will be the same size? 375

JUANA: Yes.

ANNA: So do you think this line segment we drew in is a 5 now?

JUANA: Yes. 380

ANNA: So if this is a 5, then line segment x has to be a 3.

BEN: So what about the other line segment, the one that is y? It's an 8, because if you flip the other 8 on top of it, they'll be the same size.

NOEMI: No, they won't!

I was very interested in what Ben was thinking about but didn't want to leave our discussion at this point. Luckily Arturo intervened before we got off track. 385

ARTURO: Going back to the 5 + 3, I think it could be a 5 + 3 or a 4 + 4, because both equal 8. Both could be our answer. It doesn't really matter which one because both equal 8.

I could hear others agreeing. Now what? 390

ARTURO: And it could be 1 + 7. [*More sounds of agreement.*]

NOEMI: Well, I disagree. How can this line segment we drew in be a 4 if the "front one" is a 5 and we push it in? It has to be a 3.

JUANA: It's not a 5. The line segment we drew in wasn't already there. We drew it. So it could be a 4. 395

NOEMI: It *has* to be a 5 because you're pushing it in. How could it change to a 4? It's a rectangle. Both sides need to be equal to 5. Otherwise the rectangle would look like this. [*She draws a figure to demonstrate.*] Is that a rectangle?

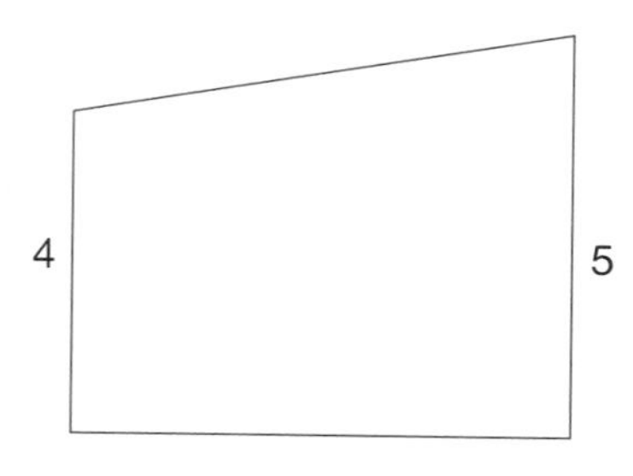

The students were very engaged in this discussion. Some were frustrated because they were convinced the answer was 3, but they were having a hard time explaining it. A few were confused totally, and a few were on the verge of understanding. I find it really hard to capture the excitement in their voices as they tried to explain their thinking to one another. Their use of logic and reasoning was impressive to me and important to them. 400

ARTURO: Well, I think I get the 5 part. So you could put the 5 on the bottom and a 4 on line segment x. 405

TEACHER: Do people agree with Arturo, that the bottom is 5 and line segment x could be a 4?

ANNA: NO! Line segment x has to be a 3 or it won't work!

ARTURO: [*mumbles under his breath*] It could be a 4!

JUAN: "Pushing it in" means moving it over to the other side. Let's push the 8 side in and try to explain it that way. 410

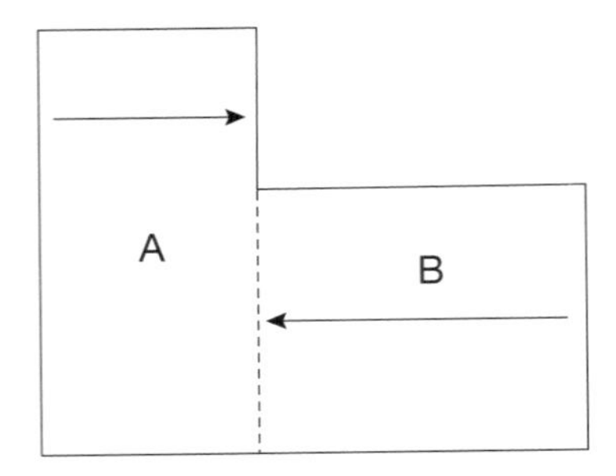

PIETRO: If you push the 8 over to segment x, and you know that the drawn in part is a 5, then how big does the other side of rectangle A have to be? Eight, right? So we know that the bottom is 5, so x has to be a 3. Because $5 + 3 = 8$.

I noticed that Julio had started to draw out the polygon on the grid paper. Maybe he will add that kind of argument to our discussion when we return to this problem. He usually offers clear explanations. I think his more concrete model might help Arturo and the others who seemed stuck. I told the class that they should continue working on this problem, and suggested that they think about the length of segment y for homework. 415

This class ended just before we reached a conclusion about segment x. With no closure on x, I wondered how they would do finding line segment y! Still, I was very pleased with the way they tried to defend their positions. They took each other seriously, listened intently, and were very patient with those who were having a hard time. They found different ways to prove their points. When "pushing in" the right side of the rectangle didn't seem to be working, Juan tried explaining it by "pushing in" the left side of the rectangle. Noemi's lopsided rectangle certainly proved her point, but not to Arturo. I don't think he was able to see the rectangles that were made when you "pushed in" either of the sides. It was a very exciting class, and I think they learned a lot about proving your point in a mathematical argument. 420 425

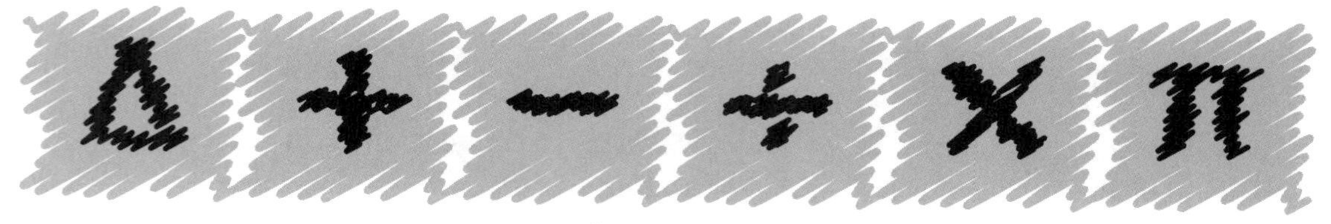

CHAPTER

8

Highlights of related research

Danielle Harrington and Marion Reynolds

As they interact with and explore their surroundings, children amass a large body of informal knowledge about shapes, even before entering school. They write and draw on rectangular pieces of paper, play with spherical balls, run their fingers over the pentagonal bolts on fire hydrants, and perhaps even notice how the angles made by the blades of a pair of scissors change size when they use them. Children may also wonder at the geometric beauty in natural forms such as butterflies, flowers, and snowflakes.

At school, the investigations of geometric objects become more formalized. Research indicates that purposeful experience with shapes is the greatest factor affecting children's understanding in geometry. In fact, the ability to reason about geometric shapes depends largely on experience. The more children interact meaningfully with shapes, the deeper and more complex their understanding becomes. This essay focuses on the early geometrical reasoning we have seen in the classroom narratives in this casebook, now considered in the light of research findings. We look at some of the mathematical issues children confront in geometry, and we examine how the nature of their thinking about geometric objects evolves.

Section 1

Reasoning about shapes

Children reason about shapes in a variety of ways. As their reasoning evolves, they move from global thinking to more analytical ways of thought. Children of elementary school age typically engage in three kinds of reasoning: reasoning by resemblance, reasoning by attributes, and reasoning by properties.

The work of Pierre van Hiele and Dina van Hiele-Geldof, Dutch mathematics educators, informs our understanding of the way children think about shapes. Research on the van Hiele theory describes different types of geometric thinking as individuals, both children and adults, gain experience and understanding in geometry (Burgher and Shaughnessy 1986; Clements and Battista 1992; Pegg and Davey 1998; van Hiele 1999). In this essay, we focus on the types of reasoning that are characteristic of elementary school children: reasoning by resemblance or appearance, reasoning by attributes or components, and reasoning by properties or relationships among attributes (Fox 2000). In progression, children's thinking moves from being dominated by visual perception to reasoning about shape in more analytical ways.

When things are unfamiliar to us, we tend not to notice the details. Instead we form an overall impression of the object. Visiting New York City for the first time, we might be struck by the size of a skyscraper and declare, "That's tall!" An architect or city planner, on the other hand, might be looking at the structural supports, the materials used, or the shape of the windows. Similarly, children with limited geometric experience see shapes as whole, or global, figures. This way of thinking is known as "reasoning by resemblance" (Fox 2000). When children pay little attention to the individual components of a shape, they are using this kind of reasoning. In talking about shapes, they might use words like fat, skinny, and long, thereby describing a shape's overall appearance.

To determine children's levels of geometric thinking, researchers frequently pose tasks that involve identifying or sorting a set of polygons. For example, children in one study were given a set of quadrilaterals and asked to put an S on each square, an R on each rectangle, a P on each parallelogram, and so forth. In another study, children were asked to sort triangles by grouping those that were alike in some way. After the children worked on these tasks, researchers interviewed the children, asking them to explain their decisions. Their responses were categorized.

At the earliest level of geometric thinking, the children answer using language from their everyday experiences (Clements and Battista 1992). An example is found in case 5, when K–1 teacher Alexandra holds up a rectangular prism and asks her students what they notice about the object. Claire replies, "It looks like a box" (p. 22). Claire's thinking is visual and linked to the overall configuration of the shape. She compares the object to something she already knows, a box.

Younger children often reason by resemblance because they tend to be more perceptually bound, less analytical, and less experienced with geometric objects. Yet older children, too, and even adults, frequently reason by resemblance when confronted with new geometric situations. In other words, visual reasoning is not simply a beginner 's way of thinking. It can be very useful, even when one is capable of other forms of reasoning.

Children using a second type of reasoning, known as "reasoning by attributes," evaluate geometric figures in terms of their specific components (Fox 2000). Attribute refers to specific geometric features of shape such as the number of sides, the length of the sides, or the size of the angles of a polygon (Hannibal 1999; Schifter 1999). Using this type of reasoning, children begin to see that a square, for instance, is made up of four "flat" parts and four "pointy" parts called sides and vertices. Students are reasoning by geometric attributes when they notice which features of shape are mathematically significant and which are not. In Bella's case 4, first grader Mark has not yet made this distinction. Asked to describe a triangular prism, Mark says it is like "half of a hill" (p. 20). He is reasoning by resemblance, not by attributes. But in case 11, Janine reports that her students described a cube by writing, "It has 6 faces, 12 edges, and 8 vertices" (p. 48). These fourth graders are calling upon the geometric components of the figure; they are "reasoning by attributes."

When children grasp which features of shape they must pay attention to in mathematical contexts, such as the number of angles or the length of sides, they experience a shift in their thinking. Using attributes as a basis for judgment, they begin to explore commonalities among shapes, such as a group of rectangles or a group of triangles. In Andrea's case 18, second-grade students consider a set of cards featuring different triangles. After a discussion about the features of these figures, the students respond in writing to the question, "What is a triangle?" Evan considers common attributes among different types of triangles he knows: "Three sides, three corners, you can turn it and it will still be a triangle" (p. 86). Evan has figured out that all triangles, regardless of their overall look, have three sides and three angles, and that orientation is an irrelevant feature in determining if a particular shape is a triangle. When children identify attributes across a variety of examples, shifting their focus away from one particular example of a shape (e.g., an equilateral triangle) to find which features are shared by all shapes in a particular category (e.g., all triangles), they are reasoning by attributes.

As children's geometric reasoning develops, they begin to identify relationships among attributes; that is, they begin to "reason by properties" (Fox 2000). Children are reasoning by properties when they are able to note, for instance, that the sides of a shape relate to each other in specific ways. They might note that in a square, all sides are equal, or opposite sides are parallel, or adjacent sides are perpendicular. In case 20, third-grade teacher Dan gives his students a homework assignment in which they must evaluate and describe shapes as parallelograms and non-parallelograms. Many children identify important attributes of parallelograms, such as the fact they have four sides, but one student, Christian, is thinking about relationships between the sides. He writes, "All the parallelograms have parallel lines . . . [and] four sides . . . The trapezoid has four lines and corners, but not all the lines are parallel" (p. 95). Reasoning by properties, Christian realizes that a trapezoid is characterized not only by the number of sides but also by how those sides relate in the shape, namely that one pair of opposite sides is parallel. Christian reasons that the trapezoid cannot be a parallelogram because it doesn't fit the requirements of a parallelogram, two pairs of parallel sides. This is an example of reasoning by properties.

Throughout the casebook, we have seen a wide range of children's reasoning in geometric situations. What these students notice and attend to differs not only from child to child but also from situation to situation. Researchers point out that students use several ways of reasoning simultaneously in a given problem-solving situation. In Rosemarie's case 3, the ideas of first grader Tessa echo the research findings on the diversity of children's reasoning. Describing a mystery block (a triangular prism), Tessa explains, "OK. It has 5 sides [faces], 6 corners . . . and it's like a ramp going down" (p. 18). In describing the overall ramp-like appearance of the prism as well as some of its attributes (number of faces and vertices), Tessa is reasoning both by resemblance and by attributes.

Reasoning on more than one level at a time is not unusual. Studies have shown significant variability in both verbal and nonverbal responses when children are presented with drawings of two-dimensional shapes (Clements et al. 1999; Lehrer, Jenkins, and Osana 1998). When asked to justify their selections of shapes for the designated category, the children in the studies hoften responded with visual "looks like" justifications (reasoning by resemblance) as well as by listing one or more attributes (reasoning by attributes). These studies suggest the levels of geometric reasoning are in fact quite fluid, varying with experience, instruction, and with the geometric task at hand. Moving through the van Hiele levels appears to be highly dependent on instruction and much less dependent, if at all, on age (Burger and Shaughnessy 1986).

Section 2

Talking about shapes

Language is one way children express the images and ideas they hold in their minds about shape. Children use their own informal language as a base on which to build a more mathematical vocabulary. As children learn about geometric objects, there is a rich interaction among their own nontechnical language, specific mathematical vocabulary, and geometrical ideas.

Putting words to spatial ideas can be challenging. In math class, children express their ideas about physical shapes and images through gestures (e.g., pointing to the face of an object), drawings, and physical actions (e.g., how they sort shapes into categories). When they articulate their ideas with spoken language, the words they use often captivate us by their originality, candor, and purity of expression. In case 6, when fourth-grade teacher Paul asks his students to describe a cone in their own words, one student describes the object as "a large circle with smaller and smaller circles on top until it reaches a point" (p. 26). This child's words reveal a particular understanding of the construction of three-dimensional solids. Through his verbal description, we have an insight into his thinking about how flat shapes stack together, face-to-face, to form a 3-D object. Indeed, children's words are windows into how they are grappling with the mathematics at hand.

Once students become conscious of geometric ideas, they begin to describe them in their own language. They express themselves freely as if describing a toy or a picture, relating new ideas to what they already know. In Mary's case 7, first and second graders describe a rectangular block as "a tree trunk," "a ladder," "a road," or "a building" (p. 32). These students are relating the elongated, rectangular form of the figure to things they know in their environment. Expressing ideas in their own natural language is a starting point in the development of children's geometric vocabularies.

Over time, children begin to attach technical terms to their thoughts. This process is revealed in students' verbal responses to tasks that involve identifying, characterizing, and sorting polygons. A third grader in one study selected four out of eight triangles and said simply they were all triangles, demonstrating her awareness of the category and of the image of triangle. A fifth grader in the same study sorted the same triangles into two sets of two triangles each, the first belonging together because both had three equal sides, and the other because both had three unequal sides. With this sorting, he demonstrated his awareness of some attributes of and mathematical terms for triangles (Burger and Shaughnessy 1986).

Even when we are aware of the correct mathematical language, there are situations in which everyday language is more useful. For instance, if the task is to describe an object to someone who is not able to see it, a triangular prism for example, initiating the description by saying the object is like a ramp or a doorstop can be very effective. On the other hand, children can use correct mathematical language without having fully developed the mathematical concepts that underlie it. For instance, in Dolores's case 17, second graders decide on the rules for being a triangle: three sides and three corners. However, they do not want to consider triangles that are "too pointy" or "too skinny" as triangles. They are still in the process of building meaning for the word triangle. The fact that they can use the vocabulary word is not enough to guarantee that they understand the ideas.

As children adopt a more mathematical vocabulary, they relate new, unfamiliar words to familiar ones. When Andrea asks her second graders to think about what makes a triangle a triangle (case 18), the children focus on the number of sides and angles of the shape. Natasha declares, "I think *tri* means three, so triangle means three" (p. 83). Natasha knows the prefix *tri* from other contexts. Combining the two parts of the word, tri and angle, Thomas then says, "three angles!" (p. 83). This ability to break down the word *triangle* into two parts they know helps the students understand the term. In these second graders, we hear echoes of the research that finds that mathematical language and the development of concepts are closely tied. When children notice, for instance, that the sides of certain triangles all have the same length, they are more ready to learn the term *equilateral triangle* and can consider why the name is appropriate (van Hiele 1999).

Children's development of mathematical vocabulary does not happen in a neat progression. Instead, there is rich, recurring interaction between nontechnical and technical language. As children solidify their ideas, they are likely to attach mathematical terms to those ideas. Conversely, children's use of precise language can help them deepen and clarify their understanding of concepts. In Janine's case 11, Ebony says, "I think . . . yeah, the square pyramid is standing on a square. I bet that's why its name is square pyramid" (p. 50). The term *square pyramid* had been applied to the object during the class. As Ebony repeated the name and analyzed it, she was able to make a statement that might serve as the beginning for a more general conclusion about the way pyramids are identified.

We notice some things without having any words for them. For example, we may recognize the faces of people we've met before without being able to describe them (van Hiele 1999). Similarly, children often develop mathematical ideas before they have specific technical language for those ideas. In fact, researchers have found evidence that some analysis of features of shape occurs at an early age, before any exposure to formal mathematics instruction (Clements and Battista 1992; Clements et al. 1999). Over time, children become increasingly aware that there is a geometric vocabulary that applies to their thinking. When Evelyn (case 1) asks her kindergartners to describe a trapezoid shape, Mitch explains, "It looks like a crystal. I don't know the name. It looks like a kite" (p. 9). Mitch seems to be aware of a specific mathematical label, separate from his own natural language, that relates to his ideas. Mitch's classmate Sandy is also cognizant of the existence of a technical vocabulary. Asked to tell what she knows about a particular shape, Sandy explains, "It is a square. It has four sides. They made that language and that's how we speak" (p. 8). Sandy's reference to the unidentified "they" who developed a mathematical way of speaking seems to highlight her perception that this language does not belong to her. Children's recognition of a special geometric language is a step in their mathematical language learning process.

As children acquire new vocabulary, they give meanings to words within a specific context (Vygotsky 1934/86). Once a child has heard a word used in a particular context, he or she associates that meaning with the word even when the context changes. It is natural for children to apply their everyday meanings for familiar terms to mathematical situations. In Molly's case 9, first and second graders compare the features of different wooden blocks. Steven and Alex begin to debate about the number of "corners" on a particular rectangular solid. Steven insists there are four corners, while Alex is convinced there are eight.

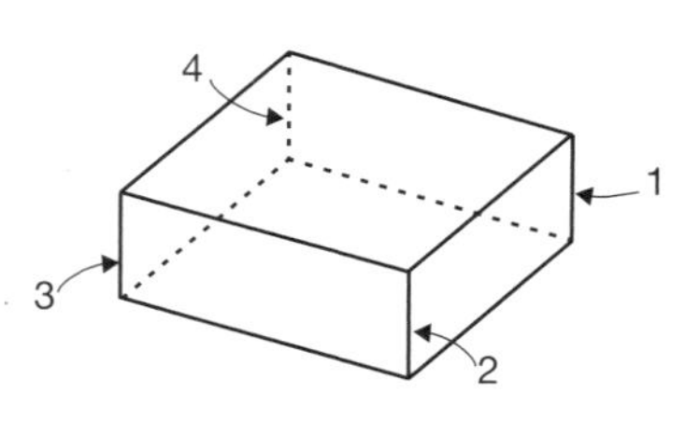

Steven's four corners

Alex's eight corners

The disagreement between the boys is rooted in the fact each has assigned a different meaning to the same everyday word *corner*; Steven is counting the vertical edges (places where the side faces intersect when the object is laid flat on a plane surface), while Alex is counting vertices (places where the edges intersect).

Children like Steven who use the familiar term *corner* to express the mathematical idea of *edge* may be thinking about the corner of a building, which, in fact, does refer to the place where the faces meet. Everyday words like corner can be ambiguous. The solution for Steven and Alex is to agree on a common meaning. When students come to realize that sharing ideas is easier if they use the same word to mean the same thing, they are, in effect, learning the need for a common mathematical vocabulary. 185 190

Section 3

Visualizing shape

Through their interactions with geometric objects, children develop mental images of what shapes look like. Children use these images as a tool for problem solving with shapes. Certain features of shape, such as symmetry or a shape's orientation in space, greatly affect children's visual interpretation of figures.

Most children have specific mental images of what shapes look like. Although these images vary from child to child, we can observe some general trends in the way children imagine shapes. They favor shapes that are symmetrical and equilateral (sides of equal length) as well as those that have a horizontal orientation. This is, after all, the way familiar shapes appear over and over again in the real world, as stop signs, yield signs, drawings of witches' hats, checkerboards, windows, doors, roof lines, bricks, and computer screens. A number of studies have investigated the kinds of mental images children tend to form. This research, which was conducted by posing shape matching and drawing tasks, suggests that students of all ages show a preference for simple and symmetrical mental images. (Bremner and Taylor 1982 and Mackay, Brazendale, and Wilson 1972, as discussed in Clements and Battista 1992.) 195 200

In fact, symmetry, orientation, and equality of side length are so important to children they seem to have internalized them as three "golden rules" by which to judge shapes. During a geometry lesson, when students are presented with shapes that do not have these characteristics, they may reject or question them. The child's mental image of a "true triangle" is often an isosceles or equilateral triangle resting on a horizontal base. As Dolores's third graders explain, even after they come to accept triangles that are oriented in different ways and that are not isosceles, the "regular" triangle will always feel "more like a triangle" (case 17, p. 78). 205 210

Orientation is a particularly salient factor affecting children's recognition of shapes. Orientation refers to the position of objects in space in relation to an external frame of reference. For many students, the way a figure is positioned becomes part of their very definition of a particular shape (Clements and Battista 1992). When shapes are moved from a horizontal orientation by being flipped (reflected) or turned (rotated), children often consider the newly positioned shape to be a completely different figure (Pegg and Davey 1998). 215

Consider, for example, a square. Many children think that a square not only must have four sides of equal length but also must sit flat on its base. In their eyes, a rotated square is not a square. When Evelyn (case 1) presents her kindergartners with a series of shapes, July examines 220 a rotated square and calls it a "diamond" (a nonmathematical term). The fact that both children and adults rename shapes when the orientation is changed reveals just how significantly orientation affects how we think about shape.

There is another aspect of orientation to consider as well. In Molly's case 2, the first and second graders were making a different argument. They reported that a triangle oriented without one side horizontal doesn't look right. As the conversation continued, they explained that such a triangle would "tip over" (p. 13). In their discussion, although they are slipping between 2-D and 3-D objects, the idea that seems important to them is stability. Researchers have reported similar findings. For instance, children's images of a right triangle were most likely to include a right triangle with a horizontal and a vertical side and least less likely to include the same kind of triangle rotated (Clements and Battista 1989).

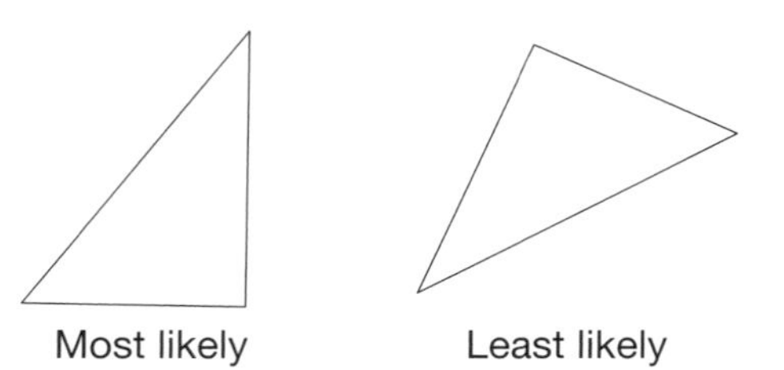

Many young children experience the delight of exploring a just emptied appliance box and seeing its configuration of squares and rectangles, corners and inverted edges, flaps and overlaps from the inside out. Additionally, most children have opened boxes to get at the cereal or a new toy inside. Through such experiences, young children construct mental images of how rectangular faces join and fold into three-dimensional boxes. Children internalize such physical and sensory experiences into coherent and flexible mental images. They store these images in their minds, refer to them, and add to them when confronted with a new geometric situation (Gravemeijer 1998).

In Anita's case 29, the third and fourth graders have been studying box construction and deconstruction for several months. Krista decides that a rectangular array cannot be folded to make a box with no overlaps, and her insight engenders a class discussion about why this is so. She explains to her classmates, "When I think of an array, the corners always get in the way" (p. 151). As she attempts to mentally fold an array into a three-dimensional box with no overlaps, she can see that the corner squares of the array have no place to go. She is able to form a mental image of a 2-D array and then manipulate that image in her mind. Krista is using visual imagery for geometric problem solving.

The role of imagery as a problem-solving tool is apparent when students work with the relationships of *sameness*, *similarity*, and *congruence*. *Sameness* is an ambiguous concept, usually understood by the context in which it appears. The word *same* may be used to describe two figures that look alike or sometimes figures that belong to the same class of shapes. *Similarity*, in contrast, describes a specific mathematical relationship between figures: Shapes are similar if one is an enlargement of the other. In judging the similarity between two shapes, students retain the image of one figure and magnify its appearance in their minds. They mentally extend side lengths and magnify whole figures, calling upon and manipulating a mental image they can compare to the shape at hand. In case 22, Josie's fourth graders are developing connections between their informal notions of sameness and the mathematical meaning of similarity as they work with plastic shapes to build similar but larger squares, rectangles, and hexagons. 255

A third relationship, *congruence between shapes*, indicates exact replication. Two shapes are congruent when one can be superimposed on the other in some way that makes an exact match. Initially, children note these relationships by sizing them up visually. That is, in judging the congruence between two figures, children get a mental image of one figure, and then manipulate this image in their mind (by rotating, flipping, or sliding it) to determine whether or not it matches the second figure. 260 265

To understand congruence, children must recognize that orientation of the figure is not a significant factor. At the beginning of Sally's case 24, second grader Mickey states that two rectangles, one oriented vertically and one oriented horizontally, are different—even though their dimensions are the same. By the end of the case, Mickey has cut out one of the shapes and moved it to cover the first. He states his conclusion: "I voted 'different' before, but now, I think I can prove they are the same" (p. 124). Mickey has used rotation to match the figures. In his case, the rotation was not mental but physical. Through such work, children form mental images of mathematical operations such as flips, slides, and rotation, and build links between these operations and the definitions of congruence and similarity. 270

Section 4

The complexities of understanding angle

Children confront a variety of mathematical issues as they examine angles. To make sense of what an angle is, they need to consider a whole range of related ideas: the connection between slant and angle, exactly what an angle comprises, what it means to measure an angle, how to see angles in geometric figures, whether or not angles are moving or static, and the connections between angle and rotation. 275

Children encounter angles every day—at street intersections, in letters of the alphabet, on fences, and in the movement of clock hands, to name a few. In their drawings, wherever straight-line segments meet or cross, angles are formed. However, even though angles can be seen everywhere in daily life, forming a complete mathematical understanding of the term is a complex process. This is partly because there are several ways to conceptualize angle. 280 285

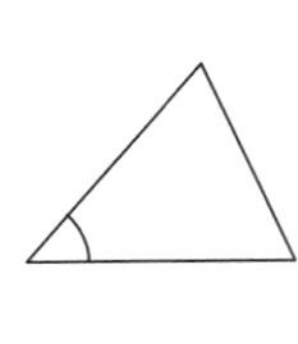

Angle in a drawn figure

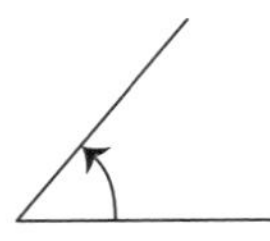

Angle as sweep or rotation

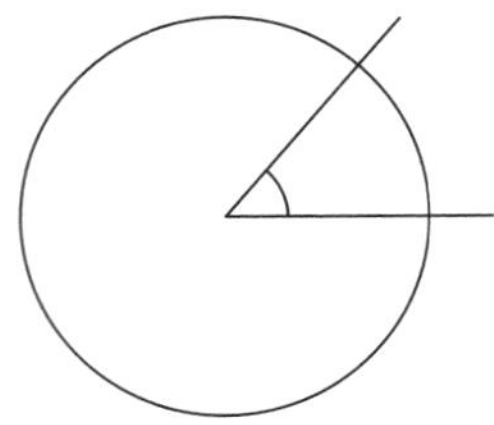

Angle as a portion of a circle

In one view, angle is the result of two lines, rays, or segments meeting or intersecting in a drawn figure; for example, consider the angles that are made by drawing a triangle or a quadrilateral. In another view, we might consider an angle as a rotation. That is, we might think of a horizontal line segment that pivots counterclockwise. Once it stops moving, the sweep of the segment, or its turn, describes an angle. In a third view, we could consider an angle in relation to a circle. For example, consider two segments (radii) drawn from the center of a circle. The two segments form an angle that can be said to represent some portion of the entire circle. While these three views of angle can be reconciled into one coherent construct, there is much to work through in that process.

The way we use the word *angle* in nonmathematical situations influences the way children think about angle. We often use it to describe the perspective from which an object is seen ("from this angle") or to describe a change of direction from a vertical or horizontal axis ("leaning at an angle"). The latter idea, "angle-as-slant," frequently crops up in children's early ideas about the nature of angle. In everyday conversation, people refer to the slant of a ramp, a playground slide, or the side of a hill as having a "steep angle." Consequently, some children come to believe that angle refers to a slanted line. In Nadia's case 13, fifth grader Alana participates in a class discussion that begins with the question, "What are angles?" Alana draws three freestanding slanted lines and announces, "These are all angles. All these lines are at an angle" (p. 61). Indeed, research corroborates that it is not uncommon for students to believe that angles are slanted lines. According to one study, fourth-grade students conceive of angle in one of three ways: as a tilted line segment, as any line segment, or as a way to indicate a direction (Clements and Battista 1990).

Other studies also show that children's ideas about angle are related to their understanding of slant. Researchers point out that children recognize slope, such as the slope of a ladder leaning against a wall, but they do not connect this idea of slant with angle. One study reports that nearly one-third of eighth graders could identify different amounts of slant in pictures of ladders, but they did not notice the related angles made by the ladder and the horizontal ground (Mitchelmore and White 2000). As they develop a fuller understanding of angle, children move from their focus on the slanted line to consider the angle formed by the slanted line and a horizontal line.

From early on in their experience with geometry, children are concerned with size. When they are reasoning by resemblance, they tend to judge a shape's size by how big or small, long or short, wide or narrow it looks. This sort of visual estimation is generally quite useful. However, when it comes to the size of angles, children's usual notions of bigness—how much space something takes up or how long or tall it is—do not apply. With everyday notions of bigness in mind, students may focus on the length of the line segments that are drawn to represent the angle, leading them to decide that angle *a* is larger than angle *b*.

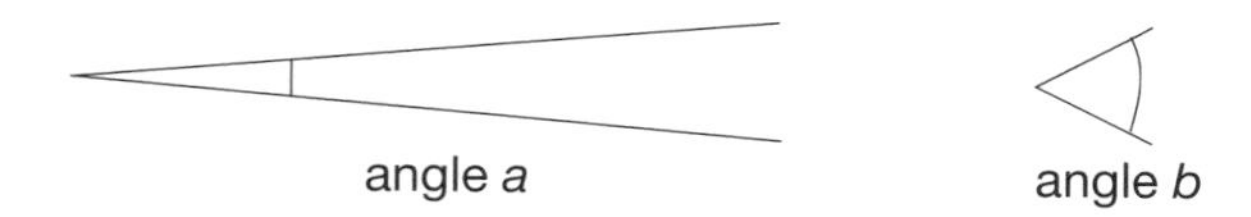

Researchers report that confounding angle measure with length, thus believing that angles with longer arms are bigger, is persistent. In fact, several studies have shown that the length of the line segments is the single most salient feature affecting children's decisions about angle size (Lehrer, Jenkins et al. 1998; Clements and Battista 1989; Clements 2003; Devichi and Munier 2013). A right angle with long line segments, such as one made by a child's outstretched arms, is judged "bigger" than a right angle made by the letter L on the page. (We may note that this is a natural assumption: a figure with longer arms is bigger than one with smaller arms.) In order for students to interpret angle size appropriately, they must modify their everyday notions of bigness. Seventh grader Casey, in Sandra's case 15, is an example of a student who has made this modification. When comparing the right angle formed by his teacher's arms with that formed by his own shorter arms, Casey understands that angles made by both sets of arms have the same measure.

Just as landmark numbers help children to anchor their investigations of the number system, landmark angles help children think about the size of angles. One frequently used landmark is the L shape of a 90° angle, which is easily compared to the shape of the angle at hand. In Jordan's case 16, one fifth grader describes this process: "Picture 90° in your head; then think, "Is it more or less?" (p. 73) In her drawing of someone seated at a desk, Ellen estimated one of the angles to have a measurement of 100° on the basis that the angle looked like it was "open" about 10° more than a right angle. She decided another angle had a measure of 45° because it seemed to be about half of a 90° angle (p. 71). Ellen based her estimates on her knowledge and mental image of right angles. Many children also create mental images of 45° or 60° angles to use as landmarks.

Another aspect of angle size emerges when children examine closed figures. Once we accept that angles can be larger than 180°, then at any point of intersection between two sides of a figure (vertex), there are two angles: one on the inside of the shape (angle 1 of the rhombus) and the other on the outside (angle 2).

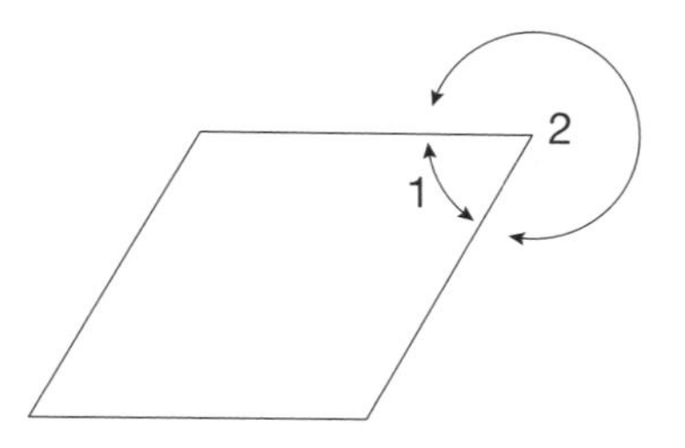

In Nadia's case 13, fifth grader Laurel questioned a classmate's statement that angles are always inside a figure. She drew a concave quadrilateral known as a chevron and suggested, "I don't think that all angles are inside a shape." Pointing to angle 2, she explains, "This is outside the shape, and there is an angle in there" (pp. 59–60). 350

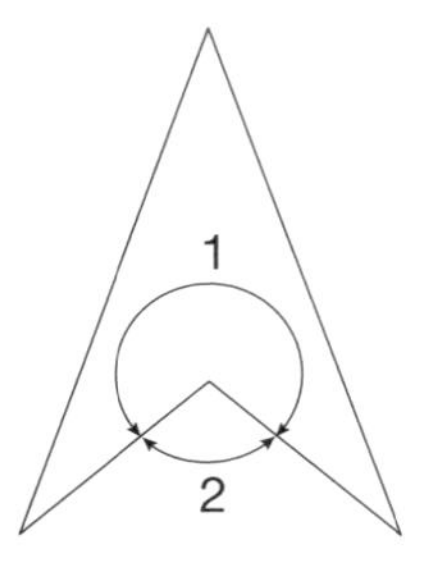

Whether Laurel believes that angle 1 is also an angle is not apparent. She has compared her mental image of an angle with the angle she sees on the outside of the chevron. Laurel may have the idea that angles must be less than 180°. As she and her classmates continue to examine such figures, they will broaden their ideas about angle to see that at any vertex there are two angles, one less than 180° and one greater. 355

In Lucy's case 14, her third graders are relating turn and movement to angle. As the students debate whether a photograph of a skater exemplifies an angle that is more than 90° or less than 90°, their discussion reveals that they are focusing on two different angles. Some note the angle formed by the skater's body itself; others envision the angle made by the movement of the skater as she bends. These students are coming to recognize that a turn or rotation creates two angles: the turn itself (angle 1), and the angle made as a result of the turn (angle 2). 360 365

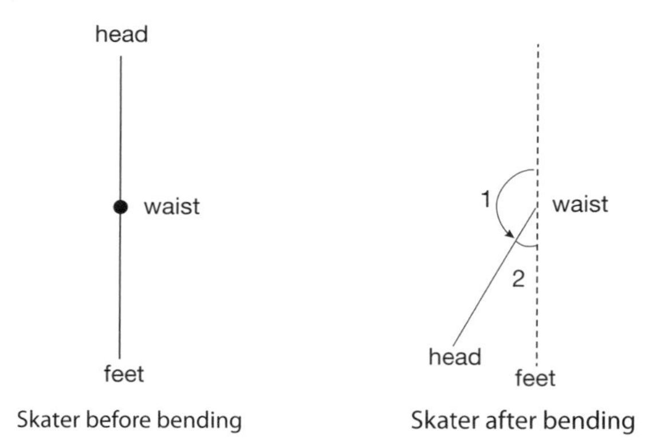

The angles in geometric figures, such as a parallelogram or triangle, can be termed static angles, while angles that move or imply motion can be termed dynamic angles. Dynamic angles can imply movement in various ways: rotation around a point (e.g., the hands of a clock), a bend or change in direction (e.g., the skater), or an opening (e.g., a door on its hinge or an elbow joint) (Mitchelmore 1998). Studies have found that children have considerable difficulty relating movement, such as turns, to angles. In a three-year study of children in the middle elementary grades, 370 researchers found that children did not call upon angle ideas when measuring, for instance, the amount of opening in a bent straw. Rather, they measured the length between its end points (Lehrer, Jenkins, et al. 1998). Understanding an angle as a degree of rotation may be difficult in part because angles have traditionally been presented to students in static contexts. As children encounter a wide variety of contexts for angles, they develop more complex understandings and their thinking about angles becomes more flexible.

Section 5

Sorting out relationships between 2-D and 3-D

Understanding the distinctions and connections between 2-D and 3-D figures is a rich and complex process. Children need to sort through the overlaps in terminology between the two dimensions, learn to interpret 2-D drawings of 3-D objects, and make sense of the way 2-D shapes come together to form a solid.

In the world outside of mathematics, we freely mix 2-D and 3-D terminology. "Roll the die to see which side it lands on," we may announce at the start of a board game, using the 2-D term *side* to mean face. "The refrigerator is an example of a rectangle," we may say, without acknowledging that we are referring to only one face of the refrigerator. Children and adults often interchange words for 2-D figures and 3-D solids as they are working with geometric objects (Oberdorf and Taylor-Cox 1999). In her K–grade 1 classroom, Alexandra (case 5) asks her students to verbalize what they notice about a variety of three-dimensional geometric solids. When this teacher holds up a triangular prism for her students to describe, Portia observes that the object "has five sides" (p. 23). Portia recognizes the five faces of the prism; she just doesn't use the mathematically correct term. In geometry, a side is a line segment that delineates the boundary of a 2-D figure (e.g., the four sides of a square). *Face* is a 3-D term that describes any of the plane (flat) surfaces that delineate the boundary of a geometric solid (e.g., the six faces of a cube).

For some students, using the mathematically correct term is not just a linguistic issue; it is a conceptual one as well. In order to meaningfully employ the term *face*, for example, children must see that what they call "sides," "top," and "bottom" are geometrically the same type of object. During shape-finding activities when children identify books as rectangles and dome-shaped clocks as circles, they are often unaware that these words describe only one face of the object. In case 25, when Eva asks her kindergartners why a tennis ball, a marble, and a wooden sphere all belong in the same category, Irene explains, "They are all circles so they go in the same group." When Eva introduces a circle drawn on a piece of paper, Carmen pipes in, "That goes with the others, because it's a circle, too" (p. 128). For Carmen, the salient feature of a circle and a sphere is that both are round. The fact that one is 2-D and the other 3-D is less important.

Some researchers have looked at the connection between the ability to draw and the ability to reason spatially (Goldsmith et al., 2016). In one longitudinal look at the development of geometric reasoning, children were asked to draw six configurations of geometric solids (a cube, a cylinder, a pyramid on top of a cube, and so forth). While drawing skill increased the most between first and second grades, subsequent verbal descriptions of their drawings did not reveal a substantial difference between the good drawers and their less-experienced peers in understanding the relationships between the 2-D and 3-D forms. Conventional views regarding the intimate relationship between drawing, spatial visualization, and many forms of geometric reasoning found little support in the results of the investigation (Lehrer, Jenkins et al. 1998; Freeman and Cox 1985).

Without question, the practice of looking at a flat drawing and calling it 3-D (e.g., "a three-dimensional drawing of a house") is contradictory. When 3-D objects are presented flat on paper, some features of the object change while others are preserved. Consider the following two ways to represent a 3-D object with a 2-D drawing. In diagram A, the angles, which are 90° in the actual cube, are drawn as either acute or obtuse; that is, angle size is not preserved. On the other hand, in diagram B, all of the angles are 90°; however, the way the faces connect and share edges is not maintained.

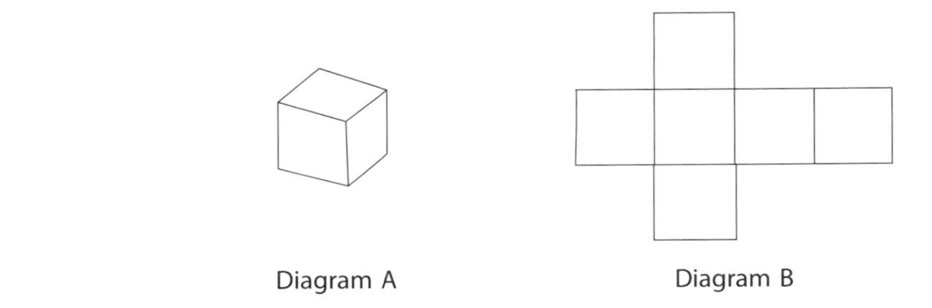

Diagram A Diagram B

In case 27 the teacher Natalie explores the question of how children maintain the 3-D relationships of objects when they try to draw them. She asks her third and fourth graders to create realistic drawings of classroom chairs. During a discussion, the children brainstorm strategies for showing depth in their drawings. One idea is to use "slanted lines" instead of perpendicular ones when drawing squares, similar to diagram A. One student, Lynda, has particular difficulty accepting this strategy. She struggles with what she knows about chairs, namely, that their seats have right angles. Lynda's image of the chair as having 90° angles is so strong that she draws the chair with only perpendicular lines and thus is unable to make a "perspective" drawing. In order to meaningfully interpret and create drawings of 3-D objects, children must become aware of artistic conventions for depicting 3-D objects, including distortion (e.g., showing a square as a rhombus) and occlusion (putting one form in front of the other) (Lehrer, Jenkins et al. 1998). They must become flexible in the way they imagine the object to look and realize that in making a 2-D drawing of a 3-D object, some features of the 3-D shapes will be preserved while others are not.

As children gain experience drawing, constructing, and discussing the structure of 3-D objects, they begin to take account of the way the separate 2-D components relate to each other to make a solid. In case 28, Olivia asks her second graders to examine and draw some geometric blocks. Brenda's drawing of a rectangular prism shows all six faces of her block in the form of a 2×3 array. She is attempting to show the relationships of the edges to each other, namely, that the side of each face touches the side of another face.

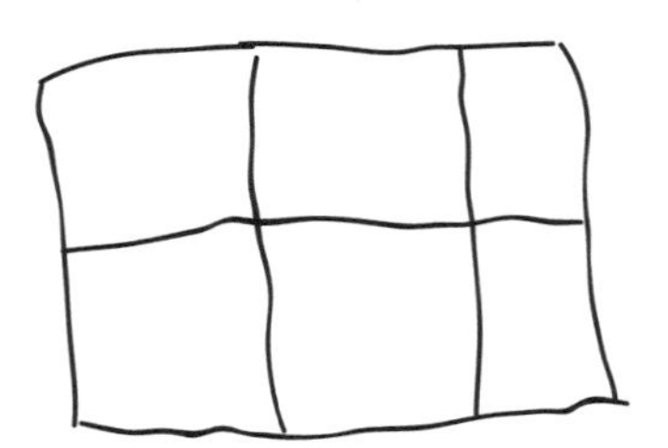

Brenda's ideas about the construction of solids are still forming. Although her drawing does not capture the three-dimensionality of her block, the fact that she shows the faces tightly touching side to side reveals her understanding of how faces connect.

Children describe the three-dimensionality of an object in a number of ways. They sometimes focus on the 2-D faces that compose an object. In Janine's case 11, fourth graders draw and describe geometric solids. Noticing that her students have ideas about the number of sides, edges, and vertices, but are not talking about how these features come together to form a solid, Janine asks her students to consider how the faces connect. When she asks Anna why a rectangular prism is three-dimensional, the girl explains, "Because if you take the top alone, or one of the sides [faces] alone, they are two-dimensional, but if you put them all together, they are three-dimensional. They are, like, 'deep'" (p. 49). Like Anna, many children focus on the exterior surface (or shell) of an object when thinking about the object's construction. They conceive of the 2-D shapes as wrapping around, folding up, or leaning against one another to compose the surface of a 3-D structure.

This *surface* view of an object contrasts with a *stacking* view that takes into account the solidness of an object. In Eva's case 25, kindergartners explore the differences between a square and a cube. Carmen suggests that you can stack papers one on top of the other to form a cube. "If you have more papers," she explains, "You can put them all together [in a pile] and you can have a block" (p. 129). One of Paul's fourth-grade students (case 6) has a similar stacking view of a sphere, which he describes as "different-size circles piled up" (p. 27). In the surface view, the sides of 2-D figures coincide; in the stacking view, the faces of the 2-D shapes coincide completely, face to face, like slices of bread in a loaf.

The notion of *edge* is important in understanding how solids are structured. In geometry, edge is a 3-D term describing the segment formed where two faces coincide. It's important for children to sort out how we apply such mathematical vocabulary to 2-D and 3-D objects. Consider the square that forms the top face of a cube. What we call the side of the square is called an edge of the cube. Isabelle's second graders (case 10) consider the notion of shared edges when they attempt to count the edges and faces of geoblock cubes. Second grader Tracy, considering how faces come together to form edges, highlights the challenge of counting those edges when she says, "Each edge that you count, and then you go to the other edges [faces] and you count them, you count the same edge again" (p. 45).

Like edges, vertices (plural of *vertex*) are also shared. *Vertex* is both a 2-D and a 3-D term. In 2-D shapes, a vertex is the point of intersection between two sides; in 3-D, it is the intersection of three or more edges.

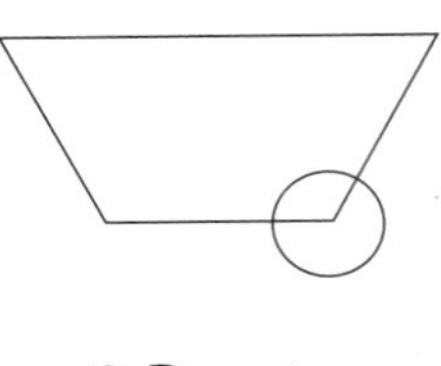

2-D vertex

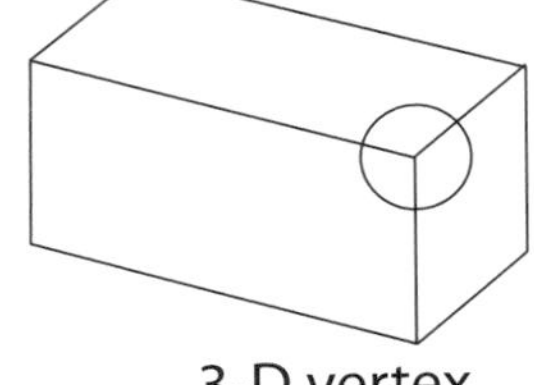

3-D vertex

In 3-D solids, children encounter the idea that the individual vertices of adjacent 2-D faces coincide neatly at one point to form one single vertex. In a rectangular prism, then, children need to coordinate all these ideas: that two sides form one edge, that three vertices form one vertex, and that three pairs of congruent rectangles form one right rectangular prism.

In building their understanding of the structure of objects, children must move between two dimensions and three dimensions (Van de Walle 1997; NCTM 2000). As they draw, build, and take apart solids, children not only examine the component parts of 3-D objects (faces, edges, vertices, and surfaces), but also see how flat surfaces come together to form an object that they can pick up, hold, and rotate with two hands.

Section 6

Building and using definitions

The geometric definitions children create are closely linked to the ways they reason about shapes in general. As students go through the process of forming and evaluating informal definitions of shapes, and as their reasoning abilities grow, their definitions become more refined.

Defining is a specific kind of mathematical activity. In the cases, we see examples of two types of defining activities: (1) making a definition for a certain category of object (e.g., deciding what makes a square a square or a triangle a triangle), and (2) creating test criterion to determine which objects belong to the category. A mathematical definition is a concise and precise list of features used to determine if a given object fits the name or not. It is the minimal list of information needed to determine a particular shape (Fox 2000; Pegg and Davey 1998). The role of definition in mathematics is thus different from definition in a nonmathematical sense. In a dictionary there is no imperative for minimalism.

How do elementary school students come to terms with the mathematical concept of definition? As they move from simply describing figures to modifying those descriptions into definitions, students grapple with definitions at a level that is tightly bound to the nature of their reasoning about shape. A less experienced student, like kindergartner Mitch in Evelyn's case 1, describes a trapezoid as "a dog food bowl" (p. 10). He finds a real-world example of the term he is trying to explain. Third grader Christian, in Dan's case 20, brings his accumulated experience with geometric objects to his informal definition of a trapezoid. "A trapezoid," Christian writes on his homework sheet, "has four lines and corners but not all the lines are parallel" (p. 95). The difference between Mitch's thinking and Christian's thinking is significant.

For children, building informal definitions often involves coming up with "rules" for shapes. Students who are reasoning by resemblance may be able to state a definition (e.g., a triangle has three sides), and at the same time explain that a scalene triangle is not a triangle because it's too "pointy." In case 18, during a game of "Guess My Rule," Andrea asks her second graders to create rules for a group of shapes which are all triangles. Despite the general consensus that triangles have three sides and three corners, Susannah cannot accept that the figures drawn on the board are all triangles. Even though the shapes fit the three-sides/three-corners rule, in her mind, the figures are not triangles. "No! I disagree!" Susannah exclaimed, "The triangles I know look like B, not like L. It's too stretched out. I still think they're just in the triangle family" (p. 84). (For Susannah, a shape can be "in the triangle family" and still not be a true triangle.)

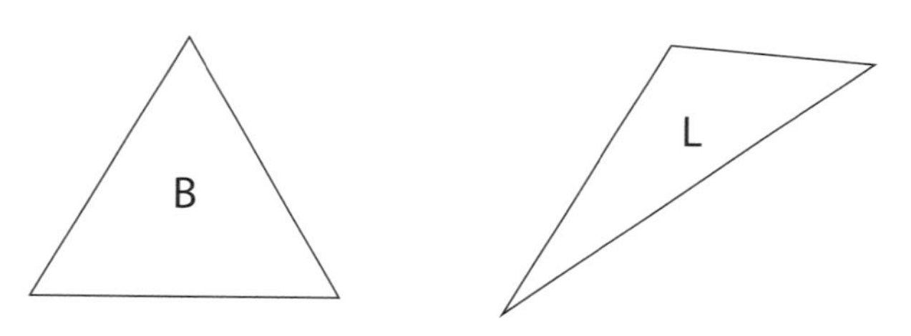

Students like Susannah, who are focused on a shape's overall appearance, often reject unfamiliar shapes because they do not correspond with their mental image associated with the defined term. They can state the definition, even generate it, but they do not apply the definition in all cases.

Similarly, when teachers ask students to construct informal definitions for figures, children who reason by resemblance will base their definitions on the way a particular shape is most often represented in the world. In Natalie's case 19, two students attempt to define a rectangle. Both base their definitions on the overall, elongated appearance they imagine rectangles to have. Maryanne characterizes a rectangle as a shape in which "two sides are longer and two sides are shorter." Finn declares, "Two squares make a rectangle" (p. 88). In the minds of Maryanne and Finn, all rectangles must be stretched out or long, so they create descriptions to fit this image.

Identifying attributes across all examples of a shape is an essential step in a child's growing ability to informally define figures. In Andrea's case 18, when second grader Thomas examines a group of figures on the board, he makes an exciting realization. "All of these shapes are triangles!" he declares (p. 84). Reasoning by attributes, Thomas is able to see the common features among the figures that make them all triangles. Conversely, his classmates are still working to 540 determine what features are characteristic of all triangles. In his description of a triangle, Jack writes, "It has 3 sides, 3 corners," but then he goes on to describe features that are dependent on the orientation of the triangle: "2 slants, 1 strate [sic] side" (p. 86). Jack seems to be thinking about one example of a triangle, an isosceles triangle sitting flat on its base. He is still figuring out which features are descriptive of all triangles and which apply only to some. While Jack 545 analyzes the figure partly in terms of its attributes (three sides, three corners), he seems to use his own mental image (that of an isosceles triangle) as a basis for his descriptions.

In their ability to characterize shapes with a variety of descriptors, students often do not see the interrelationships among attributes. For example, if a figure has three sides, it necessarily has three angles. Seeing such an interrelationship is a sophisticated understanding that typically 550 develops beyond the elementary years. Once this is understood, there is no need to list both three angles and three sides to define a triangle. But without an awareness of such relationships, elementary school students tend to generate laundry-list descriptions of shapes, naming as many features as come to mind (Clements et al. 1999; Shaughnessy and Burger 1985). In a first- and second-grade classroom (case 19), children share their ideas about squares and rectangles. When 555 the teacher asks Roberto, "What's a square?" Roberto replies, "Four sides, four corners, four angles, and it's a square" (p. 88). Mathematically, four corners (vertices) and angles are the consequence of a figure's having four sides. Therefore, listing all these attributes is not necessary. Roberto, however, is focused on the individual attributes of a square and is not yet thinking about the mathematical connections between them. 560

One important aspect of definition children need to investigate is whether or not definitions are sufficient; that is, whether they include enough information to specify a particular shape. As their reasoning about definitions grows, children discover that listing attributes alone (e.g., four sides) without regard for how they relate (e.g., congruent sides, parallel sides) may be insufficient. A definition of a particular shape is sufficient when all shapes that satisfy the condi- 565 tions of the definition are that particular shape. For example, "It has four sides" is not, alone, sufficient to define a square, because there are figures with four sides that are not squares. In Natalie's case 19, Charlie seems to be approaching an understanding of this idea of sufficiency. During a class discussion about squares and rectangles, the students brainstorm a list of attributes for each figure. For rectangles, they decide on four sides, four corners, and four angles. 570 For squares, they decide on four sides, four corners, and four angles. Pondering the two lists, Charlie notices the overlap. He calls out, "Hey, they're the exact same thing!" Charlie realizes that neither list is specific enough to fully define the shape it describes. He thinks aloud, "I'm thinking of shapes with the same definition, but they're not a square or a rectangle" (p. 89). To illustrate, he draws two quadrilaterals on the board: a chevron and a trapezoid. 575

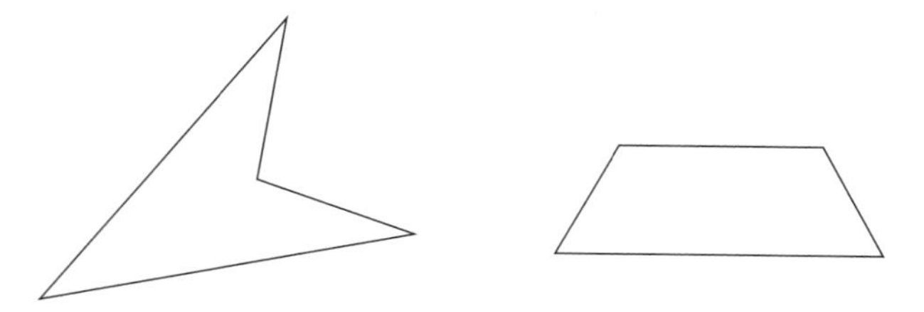

In so doing, Charlie demonstrates an ability to evaluate a definition and apply it to other shapes he knows. With his drawings, Charlie helps his classmates realize that the class list of attributes (number of sides, number of vertices) is not enough to define *square*. Paying attention to the sufficiency of descriptions is a step forward in a child's growing understanding of definition.

Creating and applying informal definitions is an appropriate goal for elementary school students. When given the opportunity to participate in the formation of informal definitions, children confront their own ideas about shapes and clarify their understanding (Keiser 2000; Koiela and Lehrer 2015). When children are asked to memorize formal definitions without having developed the reasoning to appreciate their precise, concise structure, the definition will not likely extend their understanding. Children's informal reasoning about shape and early experiences with informal definitions provide a rich base for the more formal mathematics that awaits them in the future.

Conclusion

Geometry is a rich and valuable subject for elementary school. By investigating geometry, children encounter a variety of mathematical concepts, develop reasoning skills, and often become more interested in mathematics as a whole. Geometry connects to and informs other branches of mathematics as well as other subjects, including science and art. Through their experiences with examining the features of shape, children develop a deeper appreciation of their spatial world, both human-made and natural. When encouraged to visualize, analyze, categorize, and manipulate shapes in the classroom, they are better able to understand the geometry that surrounds them in the everyday world. As children engage with geometric ideas, teachers and researchers will continue to examine the ways they make sense of these important concepts.

References

Bremner, J. G., and A. J. Taylor. "Children's Errors in Copying Angles: Perpendicular Error or Bisection Error?" *Perception* 11, no. 2 (1982): 163–171.

Burgher, W., and J. M. Shaughnessy. "Characterizing the van Hiele Levels of Development in Geometry." *Journal of Research in Mathematics Education* 17, no. 1 (1986): 31–48.

Clements, D. "Teaching and Learning Geometry." In *A Research Companion to Principles and Standards for School Mathematics*, edited by J. Kilpatrick, W. G. Martin, and D. Schifter, (pp. 151–178). Reston, Va.: National Council of Teachers of Mathematics, 2003.

Clements, D., and M. Battista. "Learning of Geometric Concepts in a Logo Environment." *Journal for Research in Mathematics Education* 20, no. 5 (1989): 450–467.

Clements, D., and M. Battista. "The Effects of Logo on Children's Conceptualizations of Angle and Polygons." *Journal for Research in Mathematics Education* 21, no. 5 (1990): 356–371.

Clements, D., and M. Battista. "Geometry and Spatial Reasoning." In *Handbook of Research on Mathematics Teaching and Learning*, edited by D. Grouws, (pp. 420–464). New York: Macmillan, 1992.

Clements, D., S. Swaminathan, M. Z. Hannibal, and J. Sarama. "Young Children's Concepts of Shape." *Journal for Research in Mathematics Education* 30, no. 2 (1999): 192–212.

Devichi, C., and V. Munier. "About the Concept of Angle in Elementary School: Misconceptions and Teaching Sequences." *Journal of Mathematical Behavior* 32, no. 1 (2013): 1–19. doi:10.1016/j.jmathb.2012.10.001

Fox, T. "Implications of Research on Children's Understanding in Geometry." *Teaching Children Mathematics* 6, no. 9 (2000): 572–576.

Freeman, N. H., and M. V. Cox, eds. *Visual Order: The Nature and Development of Pictorial Representation*. Cambridge, U.K.: Cambridge University Press, 1985.

Goldsmith, L. T., L. Hetland, C. Hoyle, and E. Winner. *Psychology of Aesthetics, Creativity, and the Arts* 10, no. 1 (2016): 56–71.

Gravemeijer, K. "From a Different Perspective: Building on Students' Informal Knowledge." In *Designing Learning Environments for Developing Understanding of Geometry and Space*, edited by R. Lehrer and D. Chazan, (pp. 45–66). Mahwah, N.J.: Lawrence Erlbaum Associates, 1998.

Hannibal, M. "Young Children's Developing Understanding of Geometric Shapes." *Teaching Children Mathematics* 5, no. 6 (1999): 353–357.

Keiser, J. "The Role of Definition." *Mathematics Teaching in the Middle School* 5, no. 8 (2000): 506–511.

Kobiela, M. and R. Lehrer. "The Codevelopment of Mathematical Concepts and the Practice of Defining." *Journal for Research in Mathematics Education* 46, no. 4 (2015): 423–454.

Lehrer, R., M. Jenkins, and H. Osana. "Longitudinal study of children's reasoning about space and geometry." In *Designing Learning Environments for Developing Understanding of Geometry and Space*, edited by R. Lehrer and D. Chazan, (pp. 137–167). Mahwah, N.J.: Lawrence Erlbaum Associates, 1998.

Mackay, C. K., A. H. Brazendale, and L. F. Wilson. "Concepts of Horizontal and Vertical: A Methodological Note." *Developmental Psychology* 7, (1972): 232–237.

Mitchelmore, M. C. "Young Students' Concepts of Turning and Angle." *Cognition and Instruction* 16, no. 3 (1998): 265–284.

Mitchelmore, M. C., and P. White. "Development of Angle Concepts by Progressive Abstraction and Generalization." *Educational Studies in Mathematics* 41, no. 3 (2000): 209–238.

National Council of Teachers of Mathematics (NCTM). "Geometry Standards for Grades 3–5." In *Principles and Standards for School Mathematics* (pp. 164–169). Reston, Va.: NCTM, 2000.

Oberdorf, C., and J. Taylor-Cox. "Shape Up!" *Teaching Children Mathematics* 5, no. 6 (1999): 340–345.

Pegg, J., and G. Davey. "Interpreting Student Understanding in Geometry: A Synthesis of Two Models." In *Designing Learning Environments for Developing Understanding of Geometry and Space*, edited by R. Lehrer and D. Chazan, (pp. 109–135). Mahwah, N.J.: Lawrence Erlbaum Associates, 1998.

Schifter, D. "Learning Geometry: Some Insights Drawn from Teacher Writing." *Teaching Children Mathematics* 5, no. 6 (1999): 360–366.

Shaughnessy, J. M., and W. Burger. "Spadework Prior to Deduction in Geometry." *Mathematics Teacher* 78, no. 6 (1985): 419–428.

Van de Walle, J. A. "Geometric Thinking and Geometric Concepts." In *Elementary and Middle School Mathematics* (pp. 342–391). New York: Addison-Wesley Longman, 1997.

van Hiele, P. "Developing Geometric Thinking through Activities that Begin with Play." *Teaching Children Mathematics* 5, no. 6 (1999): 310–316.

Vygotsky, L. S. *Thought and Language*. Cambridge, Mass.: MIT Press, 1934/1986.

ADDITIONAL RESOURCES

Battista, M. T. "The Development of Geometric and Spatial Thinking." In *Second Handbook of Research on Mathematics Teaching and Learning*, edited by F. K. Lester, (pp. 843–908). Charlotte, N.C.: Information Age, 2007.

Cooper, M. "Children's Recognition of Right Angled Triangles in Unlearned Positions." Proceedings of the Annual Conference of the International Group for the Psychology of Mathematics Education, Mexico, 1990. (ERIC Document Reproduction Service No. ED 411 138.)

Fuys, D. J., and A. K. Liebov. "Geometry and Spatial Sense." In *Research Ideas for the Classroom: Early Childhood Mathematics*, edited by Robert J. Jensen (pp. 195–222). New York: Macmillan, 1993. (Reston, Va.: NCTM, 1993.)

Goldenberg, E. P., A. Cuoco, and J. Mark. "A Role for Geometry in General Education." In *Designing Learning Environments for Developing Understanding of Geometry and Space*, edited by R. Lehrer and D. Chazan, (pp. 3–44). Mahwah, N.J.: Lawrence Erlbaum Associates, 1998.

Gravemeijer, K. "From a Different Perspective: Building on Students' Informal Knowledge." In *Designing Learning Environments for Developing Understanding of Geometry and Space*, edited by R. Lehrer and D. Chazan, (pp. 45–66). Mahwah, N.J.: Lawrence Erlbaum Associates, 1998.

Hasegawa, J. "The Concept Formation of Triangle and Quadrilateral in the Second Grade." Proceedings of the International Conference on the Psychology of Mathematics Education, Japan, 1993. (ERIC Document Reproduction Service No. ED 383 536.)

Kay, C. "Is a Square a Rectangle? The Development of First-Grade Students' Understanding of Quadrilaterals with Implications for the van Hiele Theory of the Development of Geometric Thought." Doctoral diss., University of Georgia. *Dissertation Abstracts International* 47 (1987): 2934A. (University Microfilms No. DA 8626590.)

Kosslyn, S. M. *Ghosts in the Mind's Machine: Creating and Using Images in the Brain*. New York: W. W. Norton, 1983.

Lehrer, R., C. Jacobson, G. Thoyre, V. Kemeny, D. Strom, J. Horvath, S. Gance, and M. Koehler. "Developing Understanding of Geometry and Space in the Primary Grades." In *Designing Learning Environments for Developing Understanding of Geometry and Space*, edited by R. Lehrer and D. Chazan, (pp. 169–200). Mahwah, N.J.: Lawrence Erlbaum Associates, 1998.

Mitchelmore, M. C. "The Development of Children's Concepts of Angle." Proceedings of the Annual Conference of the International Group for the Psychology of Mathematics Education, France, 1989. (ERIC Document Reproduction Service No. ED 411 141.)

Pegg, J. *Students' Understanding of Geometry: Theoretical Perspectives. Space—The First and Final Frontier.* Australia, 1992. (ERIC Document Reproduction Service No. ED 370 761.)

Piaget, J., and B. Inhelder. *The Child's Conception of Space*. New York: W. W. Norton, 1967.

Sinclair, N., M. Cirillo, and M. de Villiers. "The Learning and Teaching of Geometry." In *Compendium for Research in Mathematics Education* edited by Jinfa Cai, (pp. 457–489). Reston, Va.: National Council of Teachers of Mathematics (NCTM), 2017.